BOSTON STUDIES IN THE PHILOSOPHY OF SCIENCE
VOLUME XXIV
TECHNICS AND PRAXIS

PALLAS PAPERBACKS

Pallas Paperbacks Series is a natural outgrowth of Reidel's scholarly publishing activities in the humanities, social sciences, and hard sciences. It is designed to accommodate original works in specialized fields which, by nature of their broader applicability, deserve a larger audience and lower price than the standard academic hardback. Also to be included are books which have become modern classics in their fields, but have not yet benefitted from appearing in a more accessible edition.

Volumes appearing in Pallas will be promoted collectively and individually to appropriate markets. Since quality and low price are the two major objectives of this program, it is expected that the series will soon establish itself in campus bookstores and other suitable outlets.

PALLAS titles in print:

1. Wolff, *Surrender and Catch*
2. Fraser, *Thermodynamics in Geology*
3. Goodman, *The Structure of Appearance*
4. Schlesinger, *Religion and Scientific Method*
5. Aune, *Reason and Action*
6. Rosenberg, *Linguistic Representation*
7. Ruse, *Sociobiology: Sense or Nonsense?*
8. Loux, *Substance and Attribute*
9. Ihde, *Technics and Praxis*
10. Simon, *Models of Discovery*

A PALLAS PAPERBACK / 9

DON IHDE

TECHNICS AND PRAXIS

D. REIDEL PUBLISHING COMPANY

DORDRECHT : HOLLAND / BOSTON : U.S.A.
LONDON : ENGLAND

Library of Congress Cataloging in Publication Data

Ihde, Don, 1934–
    Technics and praxis.

    (Boston studies in the philosophy of science ; v. 24)
(Synthese library ; v. 130)
    Includes bibliographical references and index.
    1. Technology–Philosophy. I. Title. II. Series.
Q174.B67 vol. 24 [T14]   501s [601]   78-25643
ISBN 90-277-0953-X
ISBN 90-277-0954-8 pbk. (Pallas edition)

Published by D. Reidel Publishing Company,
P.O. Box 17, Dordrecht, Holland

Sold and distributed in the U.S.A., Canada, and Mexico
by D. Reidel Publishing Company, Inc.
Lincoln Building, 160 Old Derby Street, Hingham,
Mass. 02043, U.S.A.

*First published in 1979 in hardbound edition in the Reidel series
Boston Studies in the Philosophy of Science, Volume XXIV,
edited by Robert S. Cohen and Marx W. Wartofsky,
and Synthese Library, Volume 130*

All Rights Reserved
Copyright © 1979 by D. Reidel Publishing Company, Dordrecht, Holland
and copyrightholders as specified on appropriate pages within
No part of the material protected by this copyright notice may be reproduced or
utilized in any form or by any means, electronic or mechanical,
including photocopying, recording or by any informational storage and
retrieval system, without written permission from the copyright owner

Printed in The Netherlands

In Memory of Martin Heidegger

TABLE OF CONTENTS

ACKNOWLEDGMENTS ix

PREFACE: Robert S. Cohen and Marx Wartofsky xi

INTRODUCTION: Towards a Philosophy of Technology xv

DIVISION ONE / A PROGRAM IN THE PHILOSOPHY OF TECHNOLOGY

Chapter 1. The Experience of Technology: Human–Machine Relations 3

Chapter 2. A Phenomenology of Instrumentation: Perception Transformed 16

Chapter 3. A Phenomenology of Instrumentation: The Instrument as Mediator 28

Chapter 4. A Phenomenology of Instrumentation: Technics and Telos 40

DIVISION TWO / IMPLICATIONS OF TECHNOLOGY

Chapter 5. The Existential Import of Computer Technology 53

Chapter 6. Technology and the Transformation of Experience 66

Chapter 7. Vision and Objectification 82

Chapter 8. Bach to Rock, A Musical Odyssey 93

DIVISION THREE / PIONEERS IN THE PHILOSOPHY OF TECHNOLOGY

Chapter 9. Heidegger's Philosophy of Technology 103

Chapter 10. Technology and the Human: Hans Jonas 130

Chapter 11. The Secular City and the Existentialists 141

INDEX OF NAMES 151

ACKNOWLEDGMENTS

The preparation of this collection was primarily accomplished during a sabbatical, fall 1977, made possible by the State University of New York at Stony Brook and carried on in Oxford, England. Most of the essays have been previously unpublished, but almost all were originally presented at various universities and meetings. The previously unpublished essays were written between 1976–1978. Previously published essays included here and reprinted with the permission of the publishers are:

> 'The Experience of Technology: Human-Machine Relations', *Cultural Hermeneutics*, **2**, 267–279. Copyright © 1974 by D. Reidel Publishing Company, Dordrecht, Holland.

> 'Vision and Objectification', *Philosophy Today* (Celina, Ohio 45822) **17**, No. 4, Spring 1973.

> 'Bach to Rock, A Musical Odyssey', *Music and Man*, **1**, No. 1, 1973, Gordon and Breach, Science Publishers.

> 'The Secular City and the Existentialists', *Andover Newton Quarterly*, **7**, No. 4, 1967.

My interest in technology actually goes back to graduate school days in the mid sixties when located at the Massachusetts Institute of Technology and then at Southern Illinois University. At the time I was interested in the concept of work and leisure, artificial intelligence and the impact of computer technology and the role of what might be called contemporary mythology in which technology plays such a massive role.

On coming to Stony Brook in 1969 and subsequently meeting Patrick Heelan in 1970, I was stimulated by his work on a phenomenological interpretation of physics which included comments about the instrumental constitution of physical reality. It was in response to this insight that the current program in a phenomenology of instrumentation arose and I therefore must acknowledge my debt to him. I am also indebted to Lee Miller and Marshall Spector who have read and criticized parts of the present manuscript; to Ruth Cowan whose work on household technology helped stimulate new ways to think about things; and to the entire Technology, Values and Society group who are working on topics related to this project. I also wish to thank Robert S. Cohen and Marx Wartofsky who encouraged the book, and my wife, Carolyn who gives supportive criticism and grammatical help.

SUNY at Stony Brook DON IHDE

EDITORIAL PREFACE

Depending on how one construes the kinship relations, technology has been either the stepchild of philosophy or its grandfather. In either case, technology has not been taken into the bosom of the family, but has had to wait for attention, care and feeding, while the more unclear elements − science, art, politics, ethics − were being nurtured (or cleaned up).

Don Ihde puts technology in the middle of things, and develops a philosophy of technology that is at once distinctive, revealing and thought-provoking. Typically, philosophy of technology has existed at, or beyond, the margins of the philosophy of science, and therefore the question of technology has come to be posed (when it is) either by historians of technology or by social critics. The *philosophy* of technology, as analysis and critique of the concepts, methodologies, implicit epistemologies and ontologies of technological *praxis* and thought, has remained underdeveloped. When philosophy does turn its attention to the insistent presence of technology, it inevitably casts the question in one or another of the dominant modes of philosophical interpretation and reconstruction. Thus, the *logic* of technological thinking and practice has been a subject of some systematic work (e.g., in the Praxiology of Kotarbinski and Kotarbinska, among others). And the question of technology's relation to science has been posed in the framework of the nomological model of explanation in the sciences − e.g., are there 'laws' of technology; or how does technology fit within the context of justification which defines the project of a logical-empiricist philosophy of science?

But these are alternative traditions. Ihde gives a sketch of these alternatives in his introduction, noting the orientation to *praxis*, and more specifically, to technological *praxis*, to the use of instruments, or tools, in such diverse approaches as the classic theories of *technē* in Plato and Aristotle, and in such modern movements as phenomenology, Marxism and existentialism. Ihde's focus, in the end, is phenomenological. More particularly, he takes his departure from Heidegger's radical view of the primacy of technology, the initiating nature of *praxis*, with respect to science and philosophy.

Ihde's is one of the few studies which fully appreciates the centrality of technology in Heidegger's philosophy. But this is neither a study of, nor a gloss on Heidegger. Rather, Ihde originates: he starts from the contexts of *technē*, which is the human-world relation as mediated by the instrument. And in a series of carefully graded phenomenological studies, he investigates the transformations which the use of instruments and the experience of technology introduces into our 'image' of the world and to our self-image in relation to that world.

The first series of essays (Chapter 1-4) explores a phenomenology of instrumentation. Here, Ihde argues for the intentional character of *technē*, and examines the spectrum of 'instrumental intentionalities' which characterize our technological experience, (from instrumental 'transparency' to 'opacity'). Thus, he offers us an account of the origins of those epistemological and ontological errors which arise from misinterpretations or uncritical projections of our instrumental experience. Most interesting is his account of the 'objectification' or 'reification' which derives from such misconstruals of the pervasive experience that comes with the use of instruments as our probes and means of contact with the world. Modern science, in its focus on the 'monodimensionality' of what can be known by means of instrumentation, is thus subjected to an analysis which is both sympathetic and critical. Ihde's argument against 'hard technological determinism' derives from his phenomenological analysis itself, in a most interesting way.

The essays in the second part of this book vary in their focus, yet cohere as parts of a program in philosophy of technology. Thus in exploring the 'existential implications of computer technology', or in examining how our experience is transformed by technology, or in the provocative essays on the dominance of the visual model in the interpretation of the world, and of our knowledge of it, or in his essay on the differences between classical and rock music in the context of recording and listening, Ihde ramifies his earlier systematic theses in specific concrete contexts. His final essays on 'Pioneers in the Philosophy of Technology' focus on the phenomenological and existential traditions, as these develop in the philosophies of technology of Heidegger, Mounier, and Ricoeur.

The philosophy of science has finally turned its attention to philosophy of technology. More than likely, inherited parochialisms and prejudices deriving from the dominant tradition in philosophy of science will continue to shape contemporary work. All the more reason for the anti-parochial and broadening impact of Ihde's studies, which remind us of a provocative thesis of Ernst Mach: the root experience of technology as the source of our concept formations themselves. We are pleased to present this work within the Boston Studies, in all its richness of phenomenological detail and concrete analysis of *technē* and *praxis* as a contribution to the present growing discussion of philosophy of technology.

Center for Philosophy and History of Science, ROBERT S. COHEN
Boston University, MARX W. WARTOFSKY
October, 1978.

INTRODUCTION: TOWARD A PHILOSOPHY OF TECHNOLOGY

In the history and philosophy of science much talk has already occurred with respect to Thomas Kuhn's *Structure of Scientific Revolutions*. Within the language of that framework disciplines progress in terms of 'revolutions' and alternating 'normal' phases. And in any given era a discipline may be concentrated upon some 'paradigm' which in turn may undergo radical change. Were this language and framework applied to philosophy as a discipline certain anomolies would appear.

Within any given era of philosophy it might appear that rather than any single model of philosophy there would likely be a *field of philosophies* related at best by some family resemblance. Of course at some higher altitude and with the benefit of retrospective insight, it might well be that subterranean unities within the field would appear and that some generally accepted set of problems did serve 'paradigmatically'. But when a discipline does develop a paradigm with respect to either subject matter or method, it likely also finds itself with a focus such that other possible problems are overlooked or not seen at all.

As an initial task in this introduction I would like to use this language of 'paradigms', 'shifts', 'revolutions' and 'normal' phases to characterize part of the current state of philosophy. But in so doing the problem or theme which will be of central interest is *technology*. Two observations seem to be clear even at this initial state: (a) technology as a phenomenon has suddenly come into philosophic interest. There is today a rapid profusion of articles addressed to a wide variety of issures within or caused by technology. (b) But this sudden emergence of interest springs from a long and selective neglect of technology as a phenomenon. Matching today's profusion of publications on technology is a long previous silence on the part of most philosophical traditions — with some notable exceptions.

The notable exceptions are primarily those which I shall call *praxis philosophies*. Praxis philosophies, broadly defined, are those which in some way make a theory of action primary. Theory of action precedes or grounds a theory of knowledge. And it will be noted that praxis philosophies as a family have relations in widely located places within the contemporary scene.

A first, low level survey of the field of philosophies would reveal that there are a number of bloodlines, some of which have recently undergone 'revolutionary' phases and others of which are in 'normal' phases. A usual grouping of these families would probably identify (a) an Anglo-American family under the identification of *analytic philosophy*. The godfather of this family group is Logical Positivism, but its relations include a second generation which spans a spectrum which includes formalistic and constructionistic philosophies and reaches to Ordinary Language philosophy. (b) A second large grouping usually identified as Continental, includes a mixture of existentialism, phenomenology and an assortment of dialectical philosophies in the Hegel—Marx traditions. And although I shall not deal as thoroughly with them, the (c) Neo-Thomist and (d) American Pragmatist families ought to be mentioned as identifiable even though somewhat overshadowed currently by the recent 'revolutionary' phases brought about within the first two families.

A Kuhnian interpretation of this field would at least roughly identify analytic philosophy and phenomenology as 'revolutionary' with respect to earlier 19th century speculative metaphysics and might accord the same to some aspects of early twentieth century pragmatism, The other groupings would be identified as continuing 'normal' phases, although also reaction to the major thought intrusions caused by the major thinkers in the 'revolutionary' families. This, I believe, would at least be the usual view of contemporary philosophy and it calls for some thought about 'revolutions' within philosphy.

I think that a strong case can be made, within philosophy, for something like a constant potential for revolution which arises from two widely accepted and practiced ideas. Presumably philosophy as a discipline is *self-critical* and in practice philosophy changes in part by proceeding by means of *contestation*. These two ideals when taken seriously mean that nothing can stand without the threat of radical change.

Yet genuine revolutions seldom occur and long stretches of 'normal' philosophy tend to overbalance more radical perturbations. Moreover, when perturbations do occur it is often the case that some external set of events is as influential for revolution as any internal conceptual revolution. At least with respect to the analytic and phenomenological 'revolutions' of the twentieth century, it would appear that both internal and external factors were at work.

Internally, both analysis and phenomenology have claimed major revisions of the concept and task of philosophy. The 'linguistic turn' in analysis and what I shall call the 'perceptual turn' in phenomenology both refocused what was taken to be philosophy. But this internal revolution was also matched by what was at least partially perceived as a major external change as well. That external change was the at first gradual, then accelerated, emancipation and autonomy of the sciences.

Even though historically the sciences were in some deep sense offspring of philosophy, their birth, adolescence and adulthood increasingly made their autonomy obvious to thinkers at the beginning of the century. What does a parent do when the offspring have left the nest? Particularly, what does the parent do when that leaving presumably took with it a part of the general inheritance of both parent and offspring?

One alternative which was quite powerful at the turn of the century was the alternative opted for by the positivist movement and which could be called an *accommodation*. The positivist response simply relegated to the sciences any search for empirical truth and reclaimed for philosophy a narrower retirement role. This consisted largely of the seemingly modest task of dealing with purely formal and logical truth in the context of a theory of meaning. Functionally, accommodation is an alternative which appeared attractive to the sciences since they remained apparently 'free' to pursue the domains of new knowledge ceded to them by positivism. I suspect that this factor alone helps account for the often friendly reception given positivism by many in the sciences.

A second response took a quite different alternative, one which I shall call a *reclamation* strategy. Phenomenology, particularly under the original development in Husserl, responded to the autonomy of science by claiming a need to reformulate the foundations of science itself. Phenomenology even characterized itself as a type of 'new science'. Thus far from ceding the search for new knowledge to the sciences, phenomenology sought to establish a new 'scientific' vocation. Functionally such a strategy was not likely to find a completely friendly response from the newly emancipated sciences.

In both cases, however, the external events associated with the autonomy of science were matched internally by a critique of the previous grounds and habits of philosophical traditions. Both positivism and phenomenology rejected the highly speculative and systematic conceptual systems which characterized much 19th century philosophy and emerged as chastened methods which tended to pay more attention to micro-features of problems and which applied types of analyses suitable for specific problems or at most for regions as compared to the vaster metaphysics of the 19th century.

In a sense this observation points to a certain similarity between philosophy and science in practice if not in theory. Both positivism and phenomenology emerged in method as 'like' a science in focus and procedure. But there was another effect of this response as well: there began to be within philosophy a virtual preoccupation with the *philosophy of science*. This preoccupation proceeded to the extent that today the Philosophy of Science Association is the largest special interest and autonomous group within philosophy in America. Historically this is hardly surprising because the sciences are usually recognized as having made the most startling and intellectually interesting discoveries of the contemporary world. Nor should it be surprising on another count since at the very least 'natural philosophy' was parent to this now grown child and the interest of the parent remains at least paternally concerned with this lineage.

There is nothing novel in this interpretation. Indeed, analysis and phenomenology emerge from it as something like counterpart preoccupations. Those under the paradigm of the linguistic turn became preoccupied with the range of problems which centered in logic and language, while those under the perceptual turn became preoccupied with human experience in both its psychological and social dimensions. But what if one considers a new field of problems — technology as a phenomenon? Neither analysis nor phenomenology in its earliest phases seemed to be concerned with technology as such; both were concerned primarily with science.

Although this is speculation, I think it safe to say that at the origins of both analysis and phenomenology, technology would not have been thought of as being of much interest to philosophy. It may have been assumed — as I think the whole tradition has pretty much assumed until recently — that technology at best is *applied* science. But if this is assumed, it harbors a latent ontological judgment, a judgment which I shall call the philosophic preference for 'platonism'.

It might be thought that technology was not so prevasive at the turn of the century — at least with respect to the obvious complications for human existence of which we are now aware. Yet the technology of the Industrial Revolution was as massive, as polluting, as revolutionary as technology is today. And here two exceptions to the concerns of a Wittgenstein or a Husserl come to mind. Both existentialist and Marxian writers *were* concerned with technology. The Dostoievskian revolt against the rational order and the Marxian concern for alienation within a technological society even preceded the birth of analysis and phenomenology. These exceptions, to which I shall return, contained a latent emphasis upon the primacy of praxis which was as yet absent in the more central philosophic revolutions.

The dominant philosophic concerns remained those related to ontology and epistemology and for the most part excluded concern with technology as an important theme. This is perhaps understandable, given what may be thought of as kinship lines. If the sciences were offspring of philosophical parentage, it is not at all clear that technology comes from this lineage at all. The standard view that technology is applied science could at best make technology the grandchild of philosophy — but even that seems unlikely. Rather, what is more likely is that technology is that dumb brute which is to be the 'mere' instrument, tool or *slave* of science. The versions of Dr. Faustus and the Frankenstein legends all point to some creation of technology from dead or dumb matter, later brought to life through the application of theory, however bizarre or arcane.

Historically, however, technology is *older* than science in its modern guise. The implicit knowledge in *technē* was praised, but downgraded by the Greeks. But it was at least recognized as extant and powerful in the lives of humans. More recently, historians such as Lynn White have begun to argue that if the roots of modern science took life in the Rennaissance, the roots of modern technology go back at least to the Medieval period. The bloodline to philosophy, then, remains unclear and either derived at some distance or external to the central relationship altogether.

Part of the silence concerning technology comes from within philosophy itself. Philosophy usually conceives of itself more as a type of 'conceptual' engineering than as a 'material' engineering. Here there is a deeper set of relationships between science and technology as they emerge both in ancient and contemporary thought in philsosphy.

This symptomatology points to the dominance of a long 'platonistic' tradition with respect to science and technology, a tradition which with respect to science and technology turns out to be 'idealistic'. This conclusion turns upon the variable which I have called the primacy of praxis and is related only partly to the long held distinction between theory and practice.

The theory—practice distinction, however, may also be associated with a much deeper distinction, the mind—body distinction. Theory, as a set of concepts in some system of relations, is usually thought of as the product of mind, while practice often is associated with a product of body. And in the 'platonistic' tradition mind takes precedence over body. Praxis philosophies return to this tradition in a new way because the primacy of a theory of action is one which positively evaluates what I shall call the phenomena of *perception* and *embodiment*.

Contrarily, a 'platonistic' tradition is one which negatively judges or at least evaluates perception and embodiment as lower on the scale of human activity than what is presumed to be a 'pure' conceptuality. If one returns to philosophical origins, Plato's *Republic* for example, the evaluation of types or stages of knowledge is very clear. The myth of the cave and its exposition in the divided line paints a picture in which certain human capacities are simply valued more highly than other human capacities. The lowest form of knowledge, imaging (below the line) or the seeing of reflected shadows in the cave, gives way first to perceptual types of knowledge, and then to knowledge of mathematics (which is still partially perceptual in the Greek sense) and gradually rises upward to pure intuitions of the forms, presumably non-perceptual in being. True knowledge is knowledge of the forms or ideas which are supposed to be free of all taint of embodiment. Functionally this scheme downgrades both perception and embodiment.

There are, of course, cultural reasons which may be associated with this trajectory in platonism. One of these reasons lies embedded in ancient Greek religious ideas in which the dualism of body and soul is one which sees body as both the 'container' of the soul and inclined toward evil or at least away from good, whereas the soul is that ethereal and inner direction of humanity toward the Good, but a direction which is actualized by transcending 'body'.

The direction which aims toward a transcendence beyond body serves internally within platonistic theory to give a negative value to both perception and embodiment. And if this is the case, it might be thought that an opposite direction might also be developed within the Greek tradition in the form of some type of materialism. But the type of materialism which was contemporary with Plato, that of Democritus, turns out upon examination to be no more than a variant upon the same theoretical values found in platonism.

Democritus' ultimate reality is the indivisible atom — but interestingly, the atom while conceived of as 'material' is a material which is not only in fact, but in principle *unperceivable*. There is here a certain ambiguity in that the democritean atom is on one level an entity which has *reduced* perceptual qualities. Atoms have shapes, sizes and in a sense 'hardness', but lack color, texture, etc. This anticipation of a primary—secondary quality distinction, however, contains the same clearly negative evaluation of perceptual knowledge as knowledge does for Plato.

With respect to the soul, there is likewise in Democritus a trajectory which although not so radical as that of Plato, approximates it. The soul is composed of fine and ethereal atoms, unlike those grosser and coarser atoms of the body. Thus the echo of the body as 'unlike' the soul and the consequent higher evaluation of soul over body is retained even in ancient materialism. Ancient 'idealism' and ancient 'materialism' might be seen as two sides of a single theory shape.

It will be noted here that I have left out the Aristotelian tradition. This is purposeful in that it was precisely against what was taken as Aristotelian that early modern philosophy reacted. What I intend to pair here is the dominant contemporary situation with its ancient roots. With respect to science and technology, contemporarily and in the dominant traditions, it is almost as if Aristotle has dropped out.

My concern here is not directly one which speaks to this ancient distinction of mind and body so much as it is to what I think of as a rather direct analogue to this tradition which can be discerned in a distinction between science and technology relations as parallel to classical mind-body arguments and positions.

Given a strong mind–body distinction there are at least three possible dualistic alternatives: (a) A non-reductive dualism would be one which is *parallelist*. Mind and body are separate and distinct, but parallel each other and are correlated in some possible unknown fashion. The science–technology analogue could see both as distinct, but as correlated in some way.

However, non-reductive dualisms are rarely satisfying and thus one usually finds dualists turning to a reductive form of dualism and *interactionist* alternatives. Such a dualism would be one in which one factor is deemed to be originative and dominant, the other secondary and resultant. (b) Here the classical 'idealist' solution would be one which would argue that body is the result of mind or an appearance of mind. Interestingly, while this form of a reductive dualism is unpopular today in epistemology, the science–technology analogue is the dominant view of that relation. The 'idealist' view of science–technology would be one in which technology is the result of or application of science which would be the 'mind' to the 'body' of technology.

The reversal of a reductive 'idealist' dualism (c) would be a 'materialism' in which mind is thought to be emergent from or originative from body. I take it that this position or variants upon it are probably dominant in current analytic debates. But made analogous to the science–technology distinction, a 'materialist' solution in which science emerges from and is in some sense dependent upon technology is not a solution which is often posed.

There is, however, one philosopher who has both puzzled deeply over technology and who, in effect, argues for a 'materialist' solution to the science–technology relation: Martin Heidegger. His argument is one which is thoroughly steeped in the primacy of praxis and in a most radical form he argues that technology is ontologically prior to science and that science is actually the 'tool' or 'instrument' of technology.

There remains a fourth possible position with respect to the mind–body and science–technology distinction which should be mentioned. It is possible to maintain an *identity* theory of the mind—body problem. In extreme form such a theory would deny the distinction in some way and regard the distinction as a linguistic or conceptual mistake and usually, by regarding a larger whole, maintain that mind—body are identical. (I would note that most contemporary identity theories are actually forms of reductive dualism. The reduction of mind to brain states is actually a 'materialist' reductive dualism.) The analogue in the science—technology relation would be one which simply treats science—technology as a totality without distinction. In practice this position is sometimes taken in historical and social analyses of science and technology, particularly among some of the social sciences but not often within philosophy.

While this schematism is not totally satisfying, it is suggestive. And it should be clear that the dominant or at least taken-for-granted position concerning science and technology is one which is 'platonistic' and 'idealistic'. To assume that technology is applied science, that engineering is dependent upon science, that technology is made possible by science — all revolves around a presumed primacy of 'theory' over 'practice', of 'mind' over 'body'. And even among philosophers who have opted for materialist mind—body theories there remains a recalcitrant 'idealism' regarding the science—technology analogue.

In short, if there is a 'paradigm' within the dominant tradtion regarding a science—technology relation, it is one which simply takes for granted the primacy of science. What emerges from this low level scan is a recognition that technology has seldom been dealt with to the same extent as science, and that the dominant interpretation seems to be an 'idealist' one which sees technology simply as applied or instrumental with respect to science.

The result of this tradition of selectivity and interpretation may be read in the recent attempts to deal with technology within mainstream philosophy. I suggest that in some way, technology as a pervasive twentieth century phenomenon seems to exceed its causes as interpreted by platonism. This means that technology can appear as a kind of Frankenstein-phenomenon, a created 'body' which now threatens its creator. And, indeed, the strategies which currently deal with technology in much contemporary philosophy suggest precisely that one must deal with such *effects*.

One example of this strategy is the emergence of much concern over the *ethical* and *social impact* of *technology*. Programs have proliferated with respect to science, technology, society and 'values'. I shall argue that such an approach comes to technology *too late* and thus ends up dealing with symptoms rather than 'causes' of whatever dilemmas face us because of technology.

An instance of the 'value' approach can be seen in the now nearly institutional concern with medical ethics which often relate to the question of technology through its life support capacities. Life support technologies have re-opened the ancient question of a definition of life and death; technologies at the end of life are matched by those at the beginning of life with issues such as abortion and treatment of humans who would have died without technological aids. At an even deeper level, the emergence of genetic engineering within DNA research has even raised the question of life creation.

A second instance, on the political–social level, is the emergent concern with a whole range of effects made possible by social technologies. Hunger and starvation, resources management, the threat of universal destruction, possible world pollution or total environmental weather changes, have all raised questions within social and political philosophy which cannot be ignored.

However, both the ethical and the social–political questions, as urgent and as important as they are, deal with *effects* of technology rather than examine the phenomenon itself. In my view the 'value' questions come to technology too late. This can be shown to be the case even given the major assumptions of the dominant idealist and platonist view.

If one assumes that technology is an extension of science, a mere application and its instrument, then to address the effects of technology is at most to address a tertiary phenomenon. A series of relations may be formalized thusly:

$$\text{Science} \longrightarrow \text{technology} \longrightarrow \text{social effect}$$

Here the original cause is science as concept; technology is its effect or application; and the ethical or social effect is the tertiary phenomenon resultant from the series. Given this schematism, the only radical way of treating any problems which arise at the end of the series as other than symptoms would call for revision or change in the cause, in this case the *conceptual foundations of science* itself.

Of course some philosophers do precisely this. The current debates about 'value free' science, even among neo-positivist philosophers, and the attacks upon various forms of scientific reductionism are, at least within the limitations of the idealist interpretation, working at the right level. The intuition that negative results from technology lead back to possible flaws in conceptual science is at least a consistent position with respect to platonism.

If ethics comes too late, and if an interpretation of technology as a merely neutral instrument of science is inadequate, it might be that an inversion of the role of science and technology could yield some interesting suggestions. This is to posit the possibility of a different or alternative interpretation of the science–technology relation. In its analogue to the mind–body problem an inversion would clearly be a 'materialist' alternative.

What would a direct inversion have to hold? First, within the science–technology relation, it would be necessary to argue that the phenomenon is better understood as a technology–science relation. Technology becomes the origin or 'cause' of science. Thus, like a materialist version of the body–mind problem, science (mind) arises from some form of technology (body) — but it may not be immediately apparent what form this argument might take. Clearly if the body is regarded as merely mechanical, there would be problems in accounting for the excess which mind has over body, in the analogue, the conceptual excess of science over technology.

If body, however, is living and motile there is a different possibility. Motility is actualized, in this scheme, through what I have called *perception* and *embodiment*. And by this I mean that only in concrete or 'material' shapes and motions does body-technology occur. In effect, the secret of 'materialism' is the notion of *praxis*.

I shall not here outline fully what such an inverted position would have to hold but this initial projection of a 'materialist' alternative already suggests that what is called for in an interpretation of technology is a *theory of action*. And it is here that the exceptions to the philosophical neglect of technology as such become noteworthy.

Although the dominant Anglo-American traditions of philosophy fall into the 'idealist' alternative concerning technology, there is a whole family of philosophies which may be called *praxis* philosophies which have dealt with technology. This family is the family of so-called Continental philosophies, primarily existentialism, phenomenology and the dialectical tradition in its Marxian forms. Near relatives would also include some strains of Pragmatism. Each of these philosophies have a family resemblance in the way in which praxis forms the basis for other aspects of human experience.

Without elaboration, it should be obvious that Marxists within the field of philosophies *have* addressed questions to technology. The whole concept of alienation within modern industrial life is already a question of technology at least as a partial condition for social forms of life. Modes of production which play a central role in Marxist thought are social embodiments of social–political human action. Species humanity is seen to be basically actional; praxis takes historical shapes and these shapes concretely include technologies. Of course within this praxis tradition there have been widely variant attitudes toward technology. These range from the utopian views of some Marxists who see in technology the condition which can ultimately liberate humanity to distopian views which see in the centralized complexity of contemporary technology the force which ultimately threatens to reduce humanity to slavery. Moreover, the Marxist tradition is itself divided with respect to the question of the neutrality or non-neutrality of technology as such. But it remains a theory which is 'materialist' in the sense that it derives the development of conceptuality from *praxis*.

A related, but sometimes extreme form of praxis philosophy is also found in existentialism which has not neglected the question of technology. In its Sartrean version, man is what he does – humanity is act. The Sartrean project, the constant nihilation of non-being (consciousness) *is* a type of action, a praxis. Admittedly, existentialism in some of its forms was both individualistic and romantic. Thus with respect to technology one finds a positive evaluation upon what I shall call 'expressive' technologies and a negative evaluation upon organizational technologies. A tool which can be used to produce art, to express individual possibilities is 'expressive' and thus viewed positively. But any form of organizational technology which dehumanized, externally organized or alienated humans from such expressive functions would often be taken negatively as a factor in alienation. But regardless of the preference for 'expressive' technologies, existentialist writers (Dostoyevski, Berdyaev, Marcel, Jaspers) deal with technology and technological civilization to a degree seldom matched by their 'idealist' counterparts. What is important to note at this juncture, however, is that existentialism is another version of a *praxis* philosophy.

I have already noted that Heidegger, in the form of a phenomenological analysis, formulates another version of a 'materialist' interpretation of technology. It is from embodied praxis that we have technology and from this embodied form of praxis that science becomes the 'tool' of technology. Phenomenology with its rediscovery of the 'primacy of perception' and its emphasis upon concrete forms of objectification, also becomes a praxis philosophy from which a 'materialist' interpretation of technology can arise.

Such a 'materialist' inversion of the dominant 'idealist' assumptions, however, carries in its interior a whole series of yet to be examined oppositions. If a theory of action is the foundation for a theory of knowledge, if technology is the ontological source of science, if praxis is prior to concept, then no easy reconciliation between praxis philosophy and its idealist counterpart can be found. For example, one issue which is bound to arise is one which relates to the very concept of science itself. If platonists are right, then science might be a system of concepts related within some coherent whole. But if the 'materialist' alternative is correct, then science which arises from specific historical shapes of praxis is from the beginning not value free. Technology–science is from the beginning a form of action which asserts and denies various possible values. Nor is technology possible as a pure neutrality; it is a 'choice' of a possible way of being in the world. Thus ultimately, science is not a form of contemplation of the eternal forms, but is the arrangement of human social, political and individual action engaged with the material world. Such are the radically different concepts of the technology–science phenomenon to be found in the two approaches to technology. A different ontology results in a different understanding of technology, science, and the results of both within the world of human experience.

This book – actually a collection of essays on aspects of technology – is my own attempt to enter into these issues and questions. My intellectual debts are many and my sympathies, while obvious, are not entirely set. Most of my own philosophical work has been identified with American phenomenology and my sympathies are basically with what I have called praxis philosophies. But they are not so thoroughly with the praxis philosophies that they stand without doubts. For example, while I find Heidegger's work on the questions of technology perhaps the most penetrating to date, and while the radical form in which he asserts the ontological priority of technology (praxis) over science creates a new perspective upon technology as phenomenon, it is not without difficulties and problems. But I do think that this position needs to be explored in depth. And in a sense that is part of what I have done here. Indeed, I would suggest that one possible way to read the collection would be to begin with Chapter Nine, 'Heidegger's Philosophy of Technology'. Here many of my intellectual debts are revealed.

An intellectual debt, however, is not the whole of this prolegomena to philosophy of technology. I have arranged the order of chapters with a program in a phenomenology of technology. Chapter One, 'The Experience of Technology', outlines in basic form what then takes shape as a three-part instance of an analysis of technology. The instance is one which rather directly relates to questions of the relation of science and technology, *instrumentation*. My contention, although weaker than Heidegger's, is that contemporary science is *embodied* in instrumentation. Put in terms of the introduction, science as *praxis* is knowledge-gathering *activity* which only occurs by being embodied. The phenomenological variations which are explored in chapters two through four, then, carry out parts of the announced program.

Because the chapters here were originally given for various audiences and in vastly different contexts, the divisions (and the repetitions) are not always as neat as they might be. But I have addressed a series of technological phenenomena ranging from social embodiments, Chapter Five, 'The Existential Import of Computer Technology', which addresses one form of social impact through the more indirectly addressed roles of cinema, Chapter Seven, 'Vision and Objectification', and contemporary music, Chapter Eight, 'Bach to Rock, A Musical Odyssey'.

For those interested in historical and philosophical criticism, I have also included a short section on Heidegger (Chapter Nine), Jonas (Chapter Ten), Mounier and Ricoeur (Chapter Eleven).

I do not pretend that this collection is totally systematic and it certainly is not comprehensive. What it does pretend is to have done a preliminary and hopefully fundamental phenomenology of certain aspects of technology in contemporary life. In the argument which prevades the whole, that technology is non-neutral, I have tried to show the root sources for the *possibility* although not the necessity of such human experiences as alienation and fulfillment.

And although I would continue to group myself with the praxis philosophies, I am not without criticism of what often emerges as either a narrow view of technology or as an often dominantly romantic and negative tone. For example, many existentialist authors, while early aware of certain threats posed by technology, often extoll only technologies which are 'expressive'. The romanticism concerning handcraft technologies does not province a sufficient base for a critique of contemporary technological culture.

Marxists — including contemporary Critical Theory — while rightly discerning the situation of technology in terms of social structures and interests, often fail to come to grips with the foundations of technology. The very surprises that contemporary society embodied in technology has held for Marxist theory ought to evidence the partiality of this view. Yet, the focus upon social praxis does serve to demythologize the naïvete of many dominant views such as those which continue to hold to a 'value free' interpretation of science.

My hope is that this first foray into technology is suggestive. Philosophy of technology has not yet come into its own. But the philosophical inquiry into technology has begun. And precisely because philosophy of technology is clearly 'pre-paradigmatic' it also serves as an exciting possibility. Not everything has been said, nor are all the lines of argument even clear. What is most stimulating is the thought that by the very addressing of the questions of technology some of those lines of argument and 'paradigms' for discussion may be formed for the first time. That we philosophers are already very late in raising these issues seems clear to me, but we have now begun.

DIVISION ONE

A PROGRAM IN THE PHILOSOPHY
OF TECHNOLOGY

CHAPTER 1

THE EXPERIENCE OF TECHNOLOGY: HUMAN-MACHINE RELATIONS

The question of technology has been posed in many ways. That is perhaps not surprising in that the presence of technology is so pervasive. Treatments of technology span a diversity of views in philosophy as well. For example, there is a growing literature, often from the analytic tradition, which devotes itself to the question of artificial intelligence and the puzzle over whether or not machines can or could 'think'. There are others, I would term them romantic, who see in technology a growing menace to humankind's basic organic relationship to 'nature'. Still others, utopian in hope, who see quite contrarily the only possible development of humankind in terms of increasingly sophisticated technologies toward the day of a 'new Athens' in which machines finally are the slaves which allow us to develop a greater culture. Others see in certain kinds and uses of technology the potential for even greater alienation and exploitation.

However, in each of the views which criticize or extoll technology there are necessarily presuppositions concerning what happens between humans and their machines. There is at least an implicit pre-understanding of technology in order for there to be a critique or a celebration of technology. It is precisely at the level of basic presuppositions and of implicit pre-understandings that phenomenology finds its entrance into an inquiry into the question of technology. Thus, I propose to inquire into what may at first seem terribly mundane and almost too obvious for inquiry, the question of *a phenomenology of human-machine relations* as the basis for subsequent understanding of the fundamental possibilities which pose themselves within technological culture.

Clearly, phenomenology has, prior to this time, already addressed the question of technology to some extent. This is clearly the case with Heidegger both in the wider sense of posing the question of technology within its context of calculative reasoning and at the more basic level of the descriptive analysis of tools in the discussion of *Vorhandenheit* in *Being and Time*.

In a less direct way, this analysis also owes a debt to Husserl, particularly the Husserl of the *Crisis* and the analysis of the concept of the *Lebenswelt*. Within the lifeworld concept, as many commentators have pointed out, there is a dual focus. On one side, the lifeworld is regarded by a basic level of primordial experience, itself explicated by the richly implicit complexity of primary perception. On the other side, the lifeworld is also the sum of what is taken for granted, the totality of implicit beliefs and operational assumptions by which we interpret our world.

Nevertheless, in following the work of both Husserl and Heidegger toward an understanding of human-machine relations, I shall use my own terminology and adaptations from the root sources. An inter-related series of theses will be developed: (1) Although I shall show in the first section how a basically phenomenological model, whether discriminated as the Husserlian 'consciousness of ----' version of intentionality or as the more existential Heideggerian 'Analytic of Dasein' which has 'being-in-the-world' as its interpretation of intentionality, any account for the entire range of our experiences with, of, and through machines, I shall also show that experiences with machines themselves are diverse and are not simply reducible to any single primitive set. Negatively, this development will show how certain distortions of a critique and evaluation of technology become possible precisely by either an implicit or explicit choice of one type of relation as primary or paradigmatic. (2) By developing a range of the types of experience of technological artifacts, I also hope to show how pervasive our experiences of technology are and thus show its necessary impact upon the way we must conceive of the world and ultimately of ourselves. Here the underlying thesis, first in a weak sense, holds that relations with machines are non-neutral in the sense that they, by their very use, imply reflexive results for ourselves. (3) Thirdly, by displaying the variety and expanse of human-machine relations as they move toward a presumptive 'totality' the weaker thesis shows a stronger form. Human-machine relations are existential relations in which our fate and destiny are implicated, but which are subject to the very ambiguity found in all existential relations. At the same time, this existential ambiguity has its roots in the notion of the lifeworld in which the *difference* between the primordial experience of world and the sum of what is taken for granted in all conceptualized interpretations prevents any total closure into what might be called technocracy as an absolute mode of existence.

I. INTENTIONAL CORRELATIONS

Because this phenomenology may be said in some respects to 'fall between' Husserl and Heidegger, it may be wise to sketch out in brief and simplified form the basic model by which the analysis is guided. It is my understanding of a phenomenological procedure, whether interpreted as consciousness or as existential, that the core method revolves around a strategy of *correlations*.

Again, without detail or extensive review, a simple pairing of Husserl and Heidegger may be indicative:

(1) The Husserlian model of phenomenological analysis proceeds from the strict and non-separable correlation of

Ego-cogito-cogitatum.

This correlation, moreover, is the basic structure for analysis after *epochē* and its various reductions are already put into play. Note only a few central features to this model. (a) First, there is clearly a direction of analysis which begins with what is given. In the *Ideas* this is the noema, in the *Cartesian Meditations* it is the *cogitationes*. Of coures, in the widest sense, the totality of what is given with all its possibilities is the *World*. Apart from what comes second, the ego, it is important to note that what first appears is that which occurs within a world. (b) But given with world is always a way of experiencing world. Within the limits of *epochē* there is no experience without world, but neither is there any world except for an experiencing. Again, without entering the fray over the transcendental ego, what is important here is simply to note that what experiencing is and what the shape of its structure may be depends upon and is linked to what is experienced, to what is given, to noema, to world. This is to say that phenomenologically the shape of experience comes via a reflection from the world. (c) Thus we have a strict correlation in which every cogitatum is inextricably linked to the activity of cogito.

(2) This same correlation in a functional sense pervades the Heideggerian model of phenomenology. In this case, however, the mutual inter-relation of experienced and the experiencing is interpreted existentially as

Being-in-the-World.

Note now several things about the Heideggerian version of the correlation: (a) In the analysis the same 'firstness' of the World obtains. The 'worldhood of the world' comes first for the analysis. (b) Reflectively, the situation of Dasein as the being who is in the world is discerned in terms of the world and in strict correlation with the world. (c) Dasein, in this context, rather than being interpreted in a direction which lies toward the 'idealism' of Husserl, remain highly implicit and non-reified in the formula. However, the correlational model is still quite discernible.

(3) My adaptation of this direction and type of analysis leans in the direction of what might be first called a more 'American' tribal language, but which, hopefully, functions in the same way. Thus, the functional adaptation of both the Husserlian and Heideggerian models of intentionality appears here simply as:

$$\text{Human} \longrightarrow \text{World}.$$

Here the correlation of what is experienced with how it is experienced, the experiencing, is maintained. I shall add what I believe is also in Husserl and Heidegger by way of a reflexive relation as well as to explicate the implications of the correlation:

$$\text{Human} \rightleftarrows \text{World}.$$

The dashed reflexive arrow is the reflection from the World by which we may interpret or understand ourselves. Our involvements with the world are paired with the potential for reflection upon ourselves via the world in what Merleau-Ponty often called an 'arc' between ourselves and the World. Hopefully, these remarks are sufficient for the delimitation of the core paradigm of this inquiry.

II. EMBODIMENT RELATIONS

The subject of the inquiry is a preliminary phenomenology of human-machine relations. The examples I shall use will begin with quite familiar experiences. For the most part I shall focus particularly on the experienced uses of machines for reasons which will become apparent. Thus, I am not primarily interested in the beginning with dealing with the conception of, invention of, building of, or other possible human-machine relations. (This is not to say that such relations are irrelevant.)

If we begin to turn to the expanse of human-machine relations with which we are faced in daily life, what confronts us in a pre-phenomenological state is the simply vast multiplicity and extent of such relations. If, for example, I begin to take note and catalogue the sheer number of my relations with machines in a given day I may well be startled to discover how pervasive the presence of machines is. For example, trace a typical beginning of the day: I wake up to the ringing of an alarm clock; then turn to see what time it is. I soon follow this with a trip to the bathroom, perhaps my toothbrush (a simple machine) comes before, say, my weighing myself on another machine. For breakfast I employ a modern stove, a coffeepot, running water, placing the dishes in the dishwasher afterwards. To go to the university I go outdoors and start up my automobile, itself replete with various sub-machines within its totality. Arriving at the office I may use a dictaphone, a typewriter, a xerox machine or mimeograph and certainly the telephone. Meanwhile, almost unnoticed, I am surrounded by the hum of fluorescent lighting and the whir of the machine-provided heat through partially hidden vents. In fact, of course, I have indicated only a very few of the human-machine relations into which I enter on a given day — but this is enough to suggest the texture of a 'technosphere' within which we undertake our daily affairs.

Yet, while even a preliminary reflection reveals the pervasiveness of technology, for the most part our vary familiarity with machines obtrudes into the way of our understanding of those relations in a rigorous and descriptive way. Familiarity, as Heidegger so clearly points out, tends to cover over what is most significant in our relations with the World. However, I now set myself the task of addressing that taken-for-granted realm of human-machine relations and being a more specifically phenomenological set of descriptive variations in order to discern what may lie hidden within those relations.

I begin with certain simple experiences with machines and with the simple kinds of machines I can find. I pick up a pencil or a piece of chalk and begin to trace it across the desk or the blackboard. Upon a careful examination of this experience I suddenly discover that I experience the blackboard or the desk *through* the chalk — I *feel* the smoothness or the roughness of the board *at the end of the chalk.* This is, of course, also Merleau-Ponty's blind man who experiences the 'world' at the end of his cane. If I begin to be descriptively rigorous, I find I must say that what I feel is felt locally at the end of the chalk or, better, at the chalk-blackboard junction. The 'terminus' of my intentional extension into the world is on the blackboard, and I have discovered (contrary to empiricism) that touch is also a distance sense.

If I continue the reflection in terms of the phenomenological understanding of intentionality as experience within a world, I note that there is something curious about this experience. First, I clearly do not, in the case given, primarily experience the chalk as either thematic or as an object. Rather, what I experience is the blackboard and more precisely, a certain complex aspect of the blackboard's presence as texture, hardness, resistance etc. I discern that I experience the blackboard *through* the chalk, the chalk being taken into my 'self-experiencing'.

By this I mean that the chalk is only secondarily an 'object', while more primarily it is absorbed into my experiencing as an extension of myself. It is true, that the chalk is not totally absorbed in that I have what might be called an 'echo focus' in which I feel simultaneously a certain pressure at the juncture fingers/chalk with what I feel at the end of the chalk. Nevertheless, in the primary focus it is the board which I feel.

This phenomenon may now be explicated in terms of the correlation model I have already noted. However, it is important to note where the machine is placed within the correlation. In the first case above it becomes clear that the proper placing of the machine here must be upon the correlation line itself:

Human-machine ⟶ World.

The machine is 'between' me and what is experienced and is in this sense a 'means' of experience in the primary focus. Here, because the chalk is not thematized, it may be spoken of as a partial symbiotic part of the noetic act or of the experiencing of the noematic correlate in the world. This may be symbolized as follows by the introduction of parentheses:

(Human-machine) ⟶ World.

With this we have one type of human-machine relation, an experience *through* a machine. The correlational structure of intentionality remains in that I do experience something other than the machine being used and at the the same time my experiencing is extended through the machine for that intentional fulfillment. I may thus describe the chalk as having a partial *transparency relation*, between myself and what is other. And in fact, the better the machine the more 'transparency' there is. Likewise, I can use a language now which speaks of the machine as part of myself or taken into myself so far as the experience is concerned.

A series of variations would show that such relations are widespread and encompass a wide variety of human-machine relations. Furthermore, experienced uses of machines of this sort are not restricted to simple machines, but include a vast number of highly complex machines as well. For example, even driving an automobile includes relations of this type. The expert driver when parallel parking needs very little by way of visual clues to back himself into the small place — he 'feels' the very extension of himself through the car as the car becomes a symbiotic extension of his own embodiedness. The same may be said of everything from wrecking cranes to bulldozers to sailboats to airplanes. I may describe these relations as *embodiment relations*, relations in which the machine displays some kind of partial transparency in that it itself does not become objectified or thematic, but is taken into my experiencing of what is other in the World.

However, in such cases the transparency itself is enigmatic. It is clear that I do experience the board through the chalk, but it is equally clear that what is experienced is in some ways *transformed*. I do not experience the board through the chalk in the same way that I experience the board "in the flesh" with my own finger. Thus, when I compare my experience of the blackboard through the chalk and with my naked finger I may note that in both cases I get a texture with its roughness or smoothness. But with my finger I also get warmth or coolness, a spread sense of the spatiality of the board, perhaps also its dustiness or cleanness. There is a greater richness to the naked touch of the blackboard than the blackboard experienced through the chalk. I may now speak of the experiences of the blackboard through the chalk as a reduced experience when compared with my 'naked' touch of the board.

Suppose, however, I replace the chalk with a finer instrument, let us say a dentist's probe made of stainless steel with a fine pick at the end. As I trace the probe across the board I note more distinctly and clearly than before each imperfection of the board's surface. Each pock mark or crack appears through my probe in an *amplified* way, perhaps even what I neither saw nor felt with even my naked finger becomes present through the steel probe. A microscopic presence is amplified through the probe thus extending my experience of the board to a level of discernment previously unnoted.

In each of these variations in the experienced use of machines I continue to note that the embodiment relation is one in which I do experience otherness through the machine, but that the experience through the machine transforms or stands in contrast to my ordinary experience in the "flesh". This transformation takes particularly dramatic form in a sub-class of embodiment relations with which we are all familiar. I shall call these *sensory-extension-reduction* relations.

A most common such relation is almost daily experienced by most of us: the telephone call. I pick up the telephone to call a friend. Once he picks up the phone at his end and speaks I find once again that I experience him through the telephone. I speak and listen *to him* and if the connection is good the phone does not intrude primarily into our conversation. Once again there is a relative transparency. There clearly is, from one point of view, an extension of my experience — I speak to my friend from my home, overcoming geographical distance in a temporal instant. Moreover, if the technology is good the distance makes little difference. In fact, I could have the same experience of my friend whether he were in California or in Holland or next door. Note, of course, that there is a subtle and deep transformation implied here in relation to experienced spatiality. There is an almost constant 'here and now' quality of the other through the telephone, a deconstruction of certain kinds of distance. Nevertheless, the telephone 'extends' my hearing to the distant other.

However, at the same time, this extension is also a reduction in several ways. First, it is clearly a reduction of the full range of my globally sensory experience of the other. Thus, speaking through the telephone is minus the rich visual presence of the other in a face to face conversation. Imagine a variation within this reduced experience — I am speaking, rapidly, and engrossed in what I am saying, and I hear a regularly repeated 'yes', 'yes', 'oh', etc. But I do not see the yawns which accompany these words or, even worse, the facial gestures of chagrin and boredom. These remain absent and hidden from me. The other is only partially present, a quasi-presence or transformed presence. I am extended to the other, but the other is a reduced presence. This is, of course, even true of the auditory presence as well. I can recognize that it is indeed my friend who is speaking, but his speaking is 'tinny' and 'phony' when compared to the fullness of his voice while speaking face to face.

We now have two important invariants to note in terms of embodiment relations. First, in embodiment relations we do experience an otherness and in this sense the experience through a machine must be described as a partially transparent relation. Secondly and simultaneously, the experience is a transformed experience which has a difference between it and all 'face to face' or 'in the flesh' experiences. This transformation contains the possibilities, again co-implied, of both a certain extension and amplification of experience and of a reduction and transformation of experience.

Such descriptive phenomenological analyses can already be seen to hold important and extensive implications for a wide range within the experience of technology. Before passing on to a different type of relation permit me to point up just two areas of inquiry: First, there is the area of scientific instrumentation. The use of many scientific instruments includes embodiment relations, for example, the very early instruments of the telescope and the microscope.

In the use of the telescope the relation is with the moon or a planet through the telescope. And it is in the first blush of discovery that the extension and the amplification possibilities of the instrument appear. Previously unseen mountains or planetary rings now appear, are 'let be' in a new use of the Heideggerian phrase. Simultaneously, the extension and the amplification is a transformation. The universe revealed only through the the telescope and microscope retains the 'near-distance' of machine mediated experience. It remains a silent universe and still untouchable and only in recent years have we been able to adumbrate the mono-sensory silent universe through the invention of electronic machinery.

What is important to note in passing is that there is an experience of the microscopic and macroscopic universe through instruments. Scientific investigation is embodied by technology. However, it is equally important to note that such embodiment is different from the world of the naked perceptions of earthbound man.

A second more public area of importance is inquiry into media instruments such as television, cinema and radio. Again there is an experience of another and again a transformed experience, although in re-produced instances we approximate a visual-auditory text in which there is a mysterious gap between 'real time' transparency and the re-production.

What is also important in both these instances is that the *difference* between naked perception and machine mediated experience is that there occurs a subtle cross-sorting in which one becomes ambiguous in relation to the other. The world becomes a spectacle and the experiences I have in the movies become cross-sorted with those in 'real' life.

III. HERMENEUTIC RELATIONS

To this point I have been describing experiences embodied in and through machines. And I would indeed call these 'primitive' human-machine relations. They have been with humans as long as we know their history. The first club and dugout canoes entailed embodiment relations. Of course not all experiences with machines are of this type. Nor are even experiences through machines exclusive for our experiences with certain machines. If I employ a more complicated Heideggerian example and replace his hammer with my automobile I may note that so long as it functions it is taken for granted, allowing me to be embodied speedily through it. But if it begins to miss and ping while my intentional aim is disrupted, I begin to focus upon the malfunctioning motor. The automobile rapidly becomes thematized and objectified and I have an experience of the car. (I shall not pause here to reiterate the Heideggerian analysis in which he points up the context which the automobile belongs to, although that could be done. I indeed become aware of the belongingness of the automobile to the complex of repair facilities, roadways, etc. which in the automobile are far vaster and more complex than a carpenter's tool.)

What, instead, is important here is to note that we have moved from experiencing through machines to experiences of machines. However, not all these are negative. Suppose I investigate the basements of a modern university and I come upon a room filled with dials, gauges, rheostats and switches watched intently by a heating engineer. Suppose this control center monitors all the heating and cooling systems of the offices and dormitories. The engineer in the case 'reads' his dials and if one creeps up, indicating that Quad X is overheating, he merely has to turn a dial and watch to see if the heat begins to turn to normal. If it does, all right, if not, he may have to call a building manager to find out what has broken down. Here the engineer is engaged in experiences *of* a machine.

Returning to our correlational model, this experience of a machine is curious. Through the machine something (presumably) still happens elsewhere, only in this case the engineer does not experience the terminus of the intention which traverses the machine. Thus we may model the relation as follows:

$$\text{Human} \longrightarrow (\text{machine-World}).$$

His primary experiental terminus is with the machine. I shall thus call this relation a *hermeneutic relation.* There is a partial opacity between the machine and the World and thus the machine is something like a text. I may read an author, but the author is only indirectly present in the text. It is precisely in such situations that Kafkaesque possibilities may arise (imagine that the heat dial has gone awry and in fact when the engineer thinks the heat is going down it is actually going up — or better, simply imagine registrars who relate more immediately on a daily basis to computers than to students). Of course in these instances there is still a possibility of employing the difference between mediated and unmediated types of experience, the engineer could go to the dorm himself to note what was happening.

In some cases instruments probe into areas previously unknown where such checking is not at all possible, and in this case we have a genuine hermeneutic situation in which it is the hermeneut who enters the cavern to hear the saying of the oracle and we are left to his interpretation. Thus, those instruments which probe the ultramicroscopic worlds of the atom leave room for doubt as to what precisely is 'on the other side' of the machine.

I wish to point to yet another aspect of hermeneutic relations in addition to the move from the 'self-experience' absorbtion of the machine in embodiment relations to the experience with a machine in a thematized form in hermeneutic relations. In hermeneutic relations the machine becomes 'other'. But precisely because it becomes 'other' it now has different possibilities. Take, for example, a different relation with a machine, a teaching machine. When my younger daughter was nine years old she participated in an experimental program in mathematics taught via machines. In this situation the machine would pose a problem which appeared on a sheet of paper as it was typed out by the pre-programmed sequence. My daughter would then type out a solution to the problem. If the solution was correct the machine would type out something like, "All right, you've done well, go on to the next step." But if there was an error the machine would type, "No, there is something incomplete here, please go back and try again." And, if after numerous tries the solution was still incorrect the machine would type, "You must be tired for the day. Please go home and try again tomorrow". Here the appearance of 'otherness' is unmistakable. A highly reduced, but nevertheless readable 'conversation' was had between my daughter and the machine within the experience *of* a machine. This is not to say in any case that the machine has intentionality — but it is to point to the source of such pseudo-problems in the structure of the relationship itself. In relations in which machines are focal 'others' all of the ambiguity of other relations becomes a possibility. The machine is capable of anthropomorphization in terms of its 'otherness'.

We have now, two distinctive types of human-machine relations which stand in contrast: embodiment relations in which something is experienced *through* a machine, and hermeneutic relations in which the machine becomes an 'other' as a focal object of experience.

Note, too, that we have moved from one extreme to another on the continuum of the phenomenological correlation structure:

A. Embodiment relations

  (Human-machine) ⟶ World

B. Hermeneutic relations

  Human ⟶ (machine-world)

It will already be seen there that the closer to a focal thematized 'other' the machine becomes the more the significance of World must take on machine-like appearance characteristics.

IV. BACKGROUND RELATIONS

Before taking any further steps I would like to introduce yet one more set of human-machine relations. In the previous two sets of relations the use of machines were explicit and direct, either in the form of embodying a dimension of myself through a machine or in the form of confronting and being involved with a machine. However, in an increasingly more complex technological society more and more human-machine relations take on 'atmospheric' characteristics in terms of the machine background.

I return to my morning preparations. Among the catalogue of my human-machine relations stand a series of experiences among semi-automatic machines. I arise and turn up the thermostat which has been set low for the night. Unseen and barely heard, the furnace responds and in a few minutes warm air begins to circulate. As I prepare breakfast I pop some toast into a toaster which swallows the slices and in minutes returns them warm and brown. In these cases I have had a momentary relation with a machine, but in the modality of a deistic god. I have merely adjusted or started in operation the machinery which, once underway, does its own work. I neither relate through these machines, nor explicitly, except momentarily, to them. Yet at the same time I live in their midst, often not noticing their surrounding presence.

Yet their surrounding presence is almost constant. For example, in the here and now we may meet in the presence of lights, the warmth provided by our semi-automatic heating systems, and in many modern buildings in which there is a total environmental control by way of technological artifacts (none of which work well in my experience) and we may be said to be 'inside' a machine. I shall call these relations *among* machines *background relations*. And in terms of the correlation structure they may be diagramed thusly:

$$\text{Human} \rightarrow \begin{pmatrix} \text{machine} \\ \text{World} \end{pmatrix}$$

Here the illustration intends to indicate the technological texture to much of our environment; there is a 'technosphere' within which we do a good deal of our living, surrounding us in part the way technological artifacts do literally for astronauts and deep sea investigators. Explorations into the universe in environments strange and hostile to our normal being are made possible by technological cocoons we develop for ourselves. Yet, conversely, these envelopes or cocoons also give us a sense of familiarity which seems difficult to escape.

Permit one satirical illustration: The technological American plans a camping trip to the American West. He begins by encasing himself in a large camper-bus, complete with flush toilet, shower, color television, air conditioning and the lot. As he drives westward nature becomes a 'spectacle' to be viewed through the picture window of the bus, and at night he finds an instant suburbia in a crowded campground inhabited precisely by his similars, who gather in the evening to compare notes on the latest improvements to camping equipment. Here the 'technosphere' is nearly complete and has become portable for the astronaut on his own planet.

I do not mean in this facetious example to import a negative value into the discussions, because the same necessity for a technological cocoon in foreign environments makes for the possibility of ludicrous explorations of our own homeland. What I do mean to convey is the sense of partial totalization which now may be seen in terms of the continuum of the human-machine relations correlations.

V. TECHNOLOGICAL TOTALIZATION

What we have seen is the way in which human-machine relations pervade the entirety of the correlational possibilities as possibilities. Machines become, in technological culture, part of our self-experience and self-expression. They become our familiar counterparts as quasi-others, and they surround us with their presence from which we rarely escape. They become a technological texture to the World and with it they carry a presumption toward totality. In this sense, at every turn, we encounter machines existentially.

We have, of course, long thought of machines as 'extensions' of our bodies in which we existentially project ourselves into a World through machines. But with the increasing sophistication of technology, Emanual Monier, who some decades ago remarked that we should learn to think of our machines not only as extensions of our bodies, but as a development of our language, becomes more correct.

For Heideggerians, of course, this parallelism is striking and in a sense what I have done is to have linked the tool-analysis to a model which also reflects Heidegger's understanding of language as a mode of being-in-the-world. In contending that the use of machines is in the weak sense non-neutral and in the strong sense existential, I have followed an analysis which sees technological living as a way of being-in-the-world. We live and move and have our being among machines.

Clearly, the 'technosphere' contains a presumption towards totality, towards technocracy. It encompasses all dimensions of our relations. But the totality remains presumptive only. There remains the difference. Even in the face of ambiguity in which I may confuse myself with my machine world (I am "pre-programmed to learn a language" the generative linguist says, or in an earlier day, my body is a "cleverly contrived machine") there is the possibility of clarifying that difference between my meeting of the world 'in the flesh' and my meeting of the world through machines. Here is both Husserl's lifeworld in both its foci and Merleau-Ponty's measurement of the distance between the primary lifeworld and the sciences of that world which may lose their way.

But I do not wish to end on either a romantic or a pessimistic note. I only wish to indicate that a rigorous phenomenological analysis of human-machine relations poses, to my mind, the best way into an understanding of both the promises and the threats of technology. It is only *through* facing technology that we will ultimately understand it and transcend both its fascination and insidiousness.

CHAPTER 2

A PHENOMENOLOGY OF INSTRUMENTATION:

Perception Transformed

The task of this three part program concerns a descriptive, phenomenological analysis of certain aspects of the human use of 'knowledge gathering' instruments. In the inquiry I wish to take account of certain epistemological and perceptual issues which arise from the experience of such instruments.

In undertaking this task I would note, preliminarily, that what is to follow is a more detailed instance of the preceding program which described the outline of a phenomenology of human-machine relations, but that here I focus specifically upon a narrow class of such relations, those which may be found relating to 'knowledge gathering'. I am hopeful that this focus will be suggestive of certain important implications which instrumentation has for the development of science.

In the analysis I shall be largely dependent upon certain theses which arise from phenomenology as follows: (1) The use of technological artifacts, instruments in this case, may be said to 'embody' human experience in certain specific ways. It has been a commonplace to note that instruments somehow 'extend' our senses, although just in what ways and with what experiential structures may remain unnoticed. (2) However, I shall also contend that the use of such instruments — or any technological artifact — is *non-neutral*. I use this term very carefully and deliberately to suggest that there is some kind of transformation of experience in the use of instruments but I do not wish to suggest that this transformation is *ipso facto* either essentially 'good' or essentially 'bad'. (3) My major descriptive, analytic task will be to uncover more precisely and rigorously certain features of the transformation of experience involved in the use of instruments. This may be termed an investigation of the invariant features of non-neutrality. (4) I then wish to extend, at least suggestively, some of the implications which these structures of transformation may have for the epistemic situation.

By way of background and formulating a framework for the analysis I shall here first look very briefly and simply at a minimal set of methodological notions which determine the shape of the inquiry and guide the interpretation. Without going into the derivation of what phenomenologists call 'reductions', and without introducing the nuances latent within the process of 'bracketing', the core methodological notions inherent in a phenomenological analysis consist in a certain *correlational* scheme in which that which is experienced (the noema) is strictly correlated with the mode of experiencing it (noesis). I shall diagram this correlation in simple terms as one between the human experiencer and that which is experienced as:

$$\text{Human} \longrightarrow \text{World}$$

where human stands for any range of possible human experience, world stands for any possible range of what can be experienced, and the arrow stands for the relation which is usually termed 'intentional', the directedness or involvement implied in any experience of anything whatsoever. Granting that this scheme vastly oversimplifies intentionality, it will nevertheless be the guiding concept through which the analysis may proceed. There is only one more step which is needed at this most general level to specify a further implication of method. If in the first version of the diagram intentionality appears as the essential relatedness of all human experience to a field of that which is experienced, the direction is not simply one way. Not only is intentionality directed towards _____, but it also 'reflects from' that which is experienced. Indeed, phenomenologically, one wishes to distinguish strongly between any direct or 'introspective' analysis and a reflexive one. This I diagram as:

$$\text{Human} \rightleftarrows \text{World}$$

where the upper line, or 'first intentionality' is the directedness or involvement of experience with the world, and the lower or 'reflective' intentionality is the movement from that which is experienced towards the *position* from which the experience is had. Whatever falls within this correlation constitutes a proper domain for phenomenological analysis and it should be noted preliminarily that this precludes any simple talk about 'objects themselves' or, equally, 'subjects themselves', since whatever falls out in the analysis is about *relations* between experiencers and experienceds. And although this simple version of a *correlation a priori* does not yet specify all the features implied in the make-up of either the world or the human foci of the correlation, it is sufficient as a schematic framework from which to begin the analysis.

So far, now, we have a framework for investigation, a specification of a correlation as the notion of general intentionality, but we lack a specific vehicle for the investigation. In phenomenology that vehicle is composed of what might be called *variational theory*. That is, one begins by taking examples, hopefully paradigm examples, and varying them until some invariant features or structures emerge. I shall follow this process in the analysis below by selecting a finite set of paradigms to illustrate the possibilities inherent in the human use of instruments.

Ideally, within phenomenology, at first all possibilities are to be counted as 'equal'. But both in order to short-cut the long process, and to provide an internal dialectic, I shall introduce a preliminary set of distinctions which will be shown to be justified by the analysis and to be helpful in locating the types of transformation of perception which occurs with the use of instruments. I shall contrast what I shall call intentional relations 'in the flesh' with those which are 'instrument mediated'. Again, in simplest form, relations 'in the flesh' are those which may be had of things without the use of any artifact or instrument at all, while those which are mediated are relations which employ artifacts or instruments in some way.

Relating this distinction back, now, to the diagrams of intentionality, I may distinguish direct or 'in the flesh' relations as those which are simply non-technological:

$$\text{Human} \longrightarrow \text{World}$$

but those which are mediated, at least in the first instance, will include in the correlation in some way (yet to be determined) an instrument. And as the term, 'mediation', suggests, the first instance is one which places the instrument in mediational position:

$$\text{Human} \longrightarrow \text{instrument} \longrightarrow \text{World}.$$

Now I have a sufficient framework with an order of procedure to begin a concrete analysis of the human use of instruments.

EXAMPLE ONE: THE PROBE

I shall take as my first example one which I believe to be about as simple as one can imagine, the use of a probe as an 'information gathering' instrument. In this case let us use as the instance of a probe, a dentist's tool, the stainless steel, finely tipped probe he uses to feel around on our teeth. (The analysis, by the way, attends to the dentist's use, not our less pleasant end of the situation.) He is using this probe to gather information about our teeth. But let us attend to certain features of his experience in the process:

(a) First, what he feels is, of course, the texture, hardness, softness, or cracks or holes of our teeth. In this sense the probe is a *means* of his experiencing our teeth; it extends his tactile intentionality as per my previous diagram:

$$\text{Dentist} \longrightarrow \text{probe} \longrightarrow \text{tooth}$$

But a more careful look at this experienced use shows several curious things about the experience. (b) The *object* of the experience is here not the instrument, it is the tooth. The instrument is the means of the experience of the tooth. Thus in terms of the diagram, the *terminus* or fulfillment of the intention occurs at the juncture of probe–tooth. And this is, in fact, the way the tooth is experienced — the dentist feels the hardness or softness 'at the end of the probe'. (This is the dentist's equivalent of the blind man's stick noted by Merleau-Ponty in the *Phenomenology of Perception*.)

This itself is already interesting at least in contrast to the usual traditions concerning the senses. Here is touch 'at a distance'. Thus sight and sound are not the only distance senses — with the caveat that touch at a distance calls for material mediation, for some form of instrumental *'embodiment'*. (c) But note also the converse side of the sense of touch at a distance. Simultaneous to the awareness of the phenomenon: tooth as the focal entity of the experience, there is what might be called the relative *disappearance* of the probe as such. As a means of experience it *withdraws* (as Heidegger observed in his analysis of the ready-to-hand). Indeed, for the probe to function well, its disappearance must be heightened so that it becomes functionally what I shall call 'semi-transparent'. (d) This dis-appearance or withdrawal, in turn, is the way the instrument becomes the means by which 'I' can be extended beyond my bodily limit and it may be spoken of as a withdrawal into my now extended 'self-experience'. This is to say that the instrument and I as embodied become experientially a semi-symbiotic unity. I shall indicate all of these points in a modification to the diagram by placing parentheses around what is the neotic unity of this instrument use:

(Dentist–Probe) ⟶ tooth

Some general and preliminary observations about the significance of this phenomenon can already be made. First, the probe genuinely 'extends' my awareness of the world. It is the tooth I experience. There is retained in this use of the instrument, an 'instrumental realism' in which the sense of tooth remains in some way the 'same' tooth I could touch. The instrument allows me to be *embodied* at a distance, to get at the thing through the instrument.

But at the same time this genuine extension of my self-experience, is never total. There is within the experience itself a sense of *difference* which my preliminary distinction between experience 'in the flesh' and mediated experience points to. If I return to the probe example I find that while the *focal* terminus of the experience indeed occurred at the juncture probe:tooth, the fringe awareness of the instrument remained such that it never entirely disappeared. Its transparency remained partial. If I return to the experience, for example, I find that at the same time that I could feel the hardness or softness of the tooth, I had at the edge of my awareness, also a vague sense of the pressure of the instrument upon my fingers. I was clearly vaguely aware *that* I was using an instrument in the use of the instrument. I shall call this fringe awareness an echo-focus and note in terms of the diagram that the dash between myself and probe remains only semi-transparent:

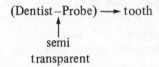

Here two suggestions pose themselves: first, it already becomes apparent that the probe in some way never fully embodies my tactile intentions and, second, there is a hint that something is being changed in the experience. With regard to this observation, there is suggested a certain ideality concerning what might be considered the 'perfect' instrument, the instrument which would reduce the dash to a 'pure transparency'. Because both of these notions are suggested in the same movement, they hold a temptation towards what I believe is a typical philosopher's category mistake regarding phenomenological accuracy and a false projection of ideality. But more of that later.

The dream of the purely transparent device does point to what may be called a *developmental telos* for technological devices. A series of variations can quickly show what is involved: If my probe is made of wood, particularly a fine hardwood, it can still give me some sense of the texture, hardness and softness of the tooth. But it remains more 'opaque' than what I sense through the stainless steel. Pushing the variation to the ludicrous, substitute a relatively soft rubber probe. Here the dampening effect is such that virtually no knowledge of the tooth is made available at all. The appropriateness of the device in these instances relates to the relative transparency which the instrument allows for our self-experience.

At the same time that the discovery of the possible extension of a bodily intentionality through the probe is suggested, and along with it the idea of a relative perfection in degree of transparency, there remains a second suggestion that within the use of the probe there occurs some kind of transformation of the experience.

I may perhaps best locate the differences within the transformation by at first comparing certain aspects of 'in the flesh' experience and the instrument mediated experience. I can, of course, feel the tooth with my finger. In this case I do get the sense of the tooth's hardness, its texture and more besides. But compared to the sense of the tooth through the probe, I now note that something is missing as well. The probe not only extended my embodiment, it *amplified* certain characteristics of the tooth. Through the probe I actually get a better sense of the hardness and softness of the tooth surfaces, a finer discrimination. The probe gives me what, compared to the fleshy finger, are *micro-features* of the tooth's surface. Thus part of the amplification of the instrument also reveals micro-features only partly available, or perhaps not at all available to my finger.

This aspect of the use of an instrument is *dramatic*, it stands out. And it is one of the sources of the fascination and amazement we experience in technological advance. (I think it should be apparent that exactly the same thing is happening through the use of a telescope which reveals to us the rings of Saturn or the mountains of the Moon, or through the microscope when it reveals the 'wee beasties' in our drinking water.)

But at the same time that the probe extends and amplifies, it *reduces* another dimension of the tooth experience. With my finger I sensed the warmth of the tooth, its wetness, etc., aspects which I did not get through the probe at all. The probe, precisely in giving me a finer discrimination related to the micro-features, 'forgot' or reduced the full range of other features sensed in my finger's touch.

But this reduction of experience is not dramatic, it is recessive. Were I not to critically re-check what is possible with the tooth with my finger, I could easily not notice or forget the rich range of tooth features available in the flesh.

I am here making two points: first, the use of the probe transforms direct perceptual experience. This is its non-neutrality. But, second, the transformation itself displays an invariant feature which I shall now call the *amplification–reduction* structure. This structure is constantly two-sided. With every amplification, there is a simultaneous and necessary reduction. And within this structure, two effects may be noted: first, the amplification tends to stand out, to be dramatic, while the reduction tends to be overlooked, or may be forgotten, particularly when the technology is truly good, when its transparency is highly enhanced. But the point is that the more enhanced the transparency, the greater the contrast between the drama of the amplification and the recessiveness of the reduction. The second effect is that the transformation also alters what may be called the 'distance' of the phenomenon being experienced. The instrument mediated entity is one which, in comparison with in the flesh relations, appears with a different perspective, its micro (or marco) features are emphasized and this is part of the transformation process itself.

I probably need not go too far to point up the implications for a scientific technology of instruments here. What I am trying to isolate are genuine *eidetic* or invariant features of the use of instruments. And while the probe is an ultimately simple example, it contains all the experience-features one can find in more complex instruments.

From this preliminary descriptive analysis of what I have called *embodiment relations*, intentional relations in which instrument mediation takes the shape indicated, certain implications for the epistemic situation may be derived: (a) First, there is a genuine relationship with the phenomenon (noema) *through* the instrument. (b) But while this is a genuine 'getting at' the object, it is a modified or non-neutral way of 'getting at' the phenomenon, marked by the eidetic structures of amplification—reduction with the consequent change in how features are presented.

However, one must overlook neither what falls out in this use of 'information gathering', nor what occurs in the process with respect to knowledge. First, it is only through the transformation which the instrument effects, that features which may be noted to be genuinely new emerge. Phenomenologically speaking, the instrument allows new noematic features to arise within the horizon of perceptual experience. Although it is the 'same' moon known to humans prior to the telescope, the very transformation which changes the distance-aspects of the moon brings into vision a new set of phenomena. Thus the drama of amplification again emerges concerning the technology, and only through the transformation does this new perspective become available. This drama and novelty are enticing within the epistemic situation, and may provide the basis for two related philosophical temptations: the first is the temptation to take the new features (micro or macro features) as 'more real' than those features which are more mundane. The second is to confuse the two dimensions of continuity and difference between ordinary or 'in the flesh' perception and instrument mediated perception.

Without going into the complexity involved, I believe that there is an implicit tendency in a certain way of taking science which falls exactly into the first temptation. 'Physical reality' may implicitly be taken to be that reality which is both instrument mediated and micro—macro structured. It is that which apparently 'lies beyond' our ordinary senses; it is that which is probed through instruments (even at the minimal level of applying simple measurements); and it is often constituted by what I shall call adding mono-dimensions together. 'Ordinary reality' is that which is constituted by direct perception in which 'middle sized' objects are experienced and whose micro—macro structures may at best be inferred. (Sellars' "scientific world" and "manifest world" distinction is one version of this division.) Such a view ignores the continuity of *instrument embodied* perception with ordinary perception.

Return for a moment to the telescopic extension of vision: what the telescope does is to reveal what from the earth-bound *position* is a micro-structure. The visual extension is, so to speak, a placing of an intentional terminus 'out there' by 'bringing close' what is there to be seen. In short, the telescope transforms space as distance and makes it near. (This is an example of what Heidegger calls *de-severance*.) But note also that in making near, the transformation is one which makes near in terms of human perceptions. Not only is the apparent surface of the moon changed (now filled with valleys and mountains) but the apparent position of the observer is changed (now as if flying overhead at a certain altitude). For every noematic change there is a corresponding noetic change. This would be the phenomenological formulation of this transformation. What is grasped, is grasped by bringing it into range but in a very deep sense, the *range itself does not change*. To be known, a phenomenon must fall into the horizon of intentionality, and fall into it in a certain way. This is what the instrument makes possible.

But at the same time, the other side of the temptation to reify instrument mediated knowledge as a greater 'reality' is the temptation to forget the invariant reduction and difference which occurs in instrument mediation. No instrument mediated relation is 'purely' transparent and any rigorous phenomenological analysis would be able to show that. The contemporary philosophical confusions concerning such issues as artificial intelligence, fantasies over 'brain transplants' and the like make, from my point of view, a serious category mistake which fails to take into account the phenomenological evidence of the reductive side of instrument mediation.

If the amplificatory side of instrument mediation is both dramatic and the condition of the possibility of bringing into range that which from ordinary position is micro and macro, then the reductive side of instrument mediation in its very tendency to recede must also be phenomenologically elicited. For this purpose I shall turn to a second example of instrument mediation, a very ordinary one.

EXAMPLE TWO: THE TELEPHONE

I deliberately chose the telephone because of its familiarity. It is easy to see that this instrument falls in certain ways into the 'information gathering' function of technology, and that it displays exactly the eidetic features of the probe in that it is to the other that I address myself. The phone becomes the means and is semi-transparent in my use of it. Its amplification is also apparent, because if the technology is good, it makes very little difference whether you are in the next room, the next state, or even in the next country, my hearing is 'extended' to you through the phone. (I would grant that the technology isn't that perfect in reducing the partial opacity of the instrument, and that one can usually tell if the call is long distance and certainly transoceanic calls are ordinarily far from the transparency of local calls.)

But I also chose the telephone because as an instrument for conversations between human beings, its reductive features are rather easy to isolate. If I return to the contrast between 'in the flesh' and instrument mediated situations, it is abundantly clear that the phone conversation vastly reduces the richness of the presence of the other. Take two trivial observations: first, the other is present primarily as voice, the full range of sensory presence which occurs ordinarily in a face to face conversation is hidden. The phone may be called a 'mono-sensory' device, it reduces the other to a voice. Note what can happen in this case — if I call and you are busy, you might go ahead and listen to me, adding an "uh-huh," and occasional "yes," etc. But all the while you might be yawning, looking distressed, or even reading your mail. I do not get the gestural, visual presence which I would get face to face. Moreover even what is focal through the phone is vastly reduced with respect to the field of richness I would get face to face. The voice, too, is reduced. Your voice, while noticeably yours, is nevertheless 'tinny' and 'phony'.

There is much more involved in the above reduction, but I shall take only the two suggestions indicated and look at their implications for 'information gathering'. The larger reduction, I noted, was a reduction of the multiple sensory dimensions of a face to face situation. This reductive feature is, in fact, a quite ordinary feature of information gathering instruments. There are few multi-sensed instruments to date and for the most part we have developed instruments which give us one thing at a time. Note, for instance, that at the beginning of the modern scientific revolution, much instrumentation arose out of the new optical technology which was developing. This was so strongly the case that not only was much of the science which developed skewed towards visual phenomena and disciplines, but it tended to overlook the other dimensions to the extent that vast regions of the universe were often actually thought to be silent. This reification of instrument mediated results has, of course, been changed with the contemporary electronics revolution which now applies auditory techniques to everything from echo-location (sonar) to sonograms of the body of unborn children.

The point I am making is a simple one, an instrument can in many cases give us only a mono-dimension of a phenomenon. And in the case of instruments probing exceedingly distant or exceedingly small phenomena, the limitation is often even greater. In fact, one physicist friend of mine contends that not only are there required ever larger and more powerful instruments to probe the smallest micro-phenomena (atom smashers for atoms, gigantic radio telescopes for distant galaxies, etc.), but that the reductive limitations increase as well. As a result, it becomes more and more the task to *infer* from reduced data. And clearly contemporary science is ingenious in this task — note what is gained from the very limited spectrograph picture. But if this reduction is forgotten, or confused with the previously noted temptation towards reification, the result is to reduce the sense of the phenomenon itself, With increased adumbrations through instruments Mars or the moon invariably turns out to be more complex than first thought. I am not only a voice at the end of the phone any more, than by analogy, a star is only its spectrographic reading, although both star and I include those aspects.

There is a second observation I wish to make about the reduction which occurs in the phone. I noted that the reduction is not only a reduction *to* a mono-dimension of the other, a voiced presence, it is also a reduction *of* the mono-dimension itself. It is here that a much more subtle form of transformation begins to emerge. It is a transformation which I shall call an *inclined focus* through the telephone. A set of simple variations will indicate what I have in mind: In the first variation I imagine a pair of lovers on the telephone. To be sure, they can speak for hours and much nuance comes through in the modulated tones of the conversation. But all of this is usually experienced as a very poor substitute for being in each other's arms (although much better than not being able to converse and respond — hence better than a long delayed letter). The fleshly richness of the face to face is lacking. Variation two is one in which I am conducting interviews over the telephone and while I find very dramatically that I get a good sense of the person compared to what appeared on a mere vita and in letters, yet I still find that a whole range of nuance is missing — we must have the candidate here in person to perceive the whole. But in variation three, I imagine conducting a series of business operations through the telephone. I inform my broker that I wish such and such a stock or I ask my provost to approve a delayed travel request. In the latter variation I find no need for more, it is an exchange of information concerning practical matters.

Now, I do not mean by these variations to suggest that the telephone is inappropriate in any of the cases, but I do mean to suggest that the *ratio* of amplification to reduction is more dramatic in some than in others. The phone is sufficient in some cases, while relatively inadequate in others and what is of importance is to discern towards what kinds of situations the sufficiency increases.

I should like to suggest that what arises from this example, is the inclination of this particular technology to certain kinds of pragmatic and daily affairs and particularly the conveying of 'information'. This is to say that this particular technology is inclined to purveying 'information'. But this, too, is a kind of reduction in the sense that 'information' is only one dimension of the total richness of human experience.

In some respects this is only to note that the telephone is, after all, more appropriate for some functions than others, and in that regard it simply embodies some social purpose. But in other respects I am suggesting that insofar as technologies are non-neutral they have a reflexive amplificatory—reductive effect as well.

I shall not develop this suggestion further except to note that the inclination of much contemporary technology towards information as such refracts back upon us in terms of the very ways in which we understand both world and ourselves. But to make this point calls for far more than the first step into what I have called embodiment relations.

CONCLUDING REMARKS

What I have developed here is but a first part of the analysis of the use of instruments and the ways in which they function intentionally. I have tried to show how the use of instruments does indeed gather information and even bring within the horizon of human experience that which is genuinely new. This gain in knowledge, however, occurs within the context of certain transformational features which make up the non-neutrality of instrument mediation.

I have tried to isolate certain of the essential features of the transformation by elucidating the amplification—reduction structure with some of its attendant effects and some of its dangers for misunderstanding the epistemic situation.

I have implicitly suggested that a rigorous phenomenological analysis of such formations of intentionality provides a critical tool not only for understanding what occurs in the human experience of instruments, but foreguards against certain temptations either to reify reductive results or to claim either too little or too much for what is increasingly a technological way of being in the world.

CHAPTER 3

A PHENOMENOLOGY OF INSTRUMENTATION:

The Instrument as Mediator

Martin Heidegger, as early as *Being and Time*, pointed to the semi-transparency equipment attains in use. His analysis of tools pointed out that in use the tool "withdraws" because what is focal is the "work". At the same time, he allowed for the disappearance of such transparency when the tool or instrument breaks down, malfunctions or is missing. In such cases the instrument obtrudes into the telic aim of the user and the functioning of the intentional relation ceases. In Heidegger's analysis the instrument becomes a "thing" and loses its transparency.

In spite of the phenomenological correctness of Heidegger's analysis, the *negative* way in which the instrument emerges from transparency in use in his analysis casts a sense of disvalue upon any positive thematization of an instrument.

In this essay I shall attempt to show that there is another way in which the instrument ceases to be semi-transparent, by demonstrating that there is a continuum of intentional instrumental uses such that what may be called instrumental *opacity* takes on positive phenomenological characteristics while also continuing the value of instrumental investigation which brings the previously hidden or unsuspected features of the world into view.

My strategy remains thoroughly phenomenological in that I shall employ a graded series of variations which display the continuum from one end where pure transparency remains the ideal towards the other end where instrumental opacity begins to appear in use but with the positivity associated with all phenomenological appearances.

Additionally, this strategy will begin to lay the groundwork for what I shall later call telic directions which are latent within the development of contemporary, sophisticated instrumentation. These directions point to different trajectories, one of which returns to the lived world of perception and which, even in contemporary instrumentation, aims at the maximum recovery of transparency. But the other direction is one which develops what may metaphorically be called 'instrumental intentionality.'

A. THE CONTINUUM OF INSTRUMENTAL VARIATIONS

I began the phenomenology of instrumentation with a preliminary examination of embodiment relations, human-instrument-world relations in which the mediating position of the instrument directly embodies human perceptions and motions, and in which the instrument phenomenologically might be said to be semi-transparent. Here the ideal of a pure transparency, never attained, could occur, although the highest degree of human-instrument symbiosis is precisely what characterizes such uses of instruments. The formalism, (Human—instrument) ⟶ World, symbolizes the instrument mediation as one of an embodiment position.

Such experienced uses of instruments, I shall now show, occupies one end of a graded continuum of uses, the other end of which culminates in a qualitatively different relation. In the descriptive analysis to follow, however, I shall remain limited to relations which retain the intentional form, Human—instrument—World, in which the instrument retains its mediational position 'between' the observer and that which is observed. I shall also remain within the limits of a visual—optical development in which all the instruments illustrating the continuum produce some type of visual result.

It will be seen that the latent secrets of instrumentally mediated knowledge lie in the very invariant structures which, however minimally, transform perception. The amplification-reduction structure, in these examples, retains its outline. It remains now to take note of the optical, instrumental continuum through a partially developed analysis of instrumental variations:

1. The use of a magnifying glass is directly similar to the use of a dentist's probe (discussed in Chapter 1) in that the visual features it shows are discerned directly and nearly effortlessly *through* the glass. Moreover, the magnification of micro features, although still very continuous with what is seen with the naked eye, begins to make the possibilities of the amplification—reduction structure apparent. The selectivity of this instrument is such that vision is embodied, not discontinuously, through the instrument and the first stages of the micro-world begin to come into the visual horizon. (I shall, in the following examples, concentrate upon the dramatic, amplificational side of the mediation, although its synonomous reductivity also remains constant even if harder to locate without a more thorough analysis. The very change of scale in magnification transforms 'in the flesh' perception and is a clue to certain aspects of reduction.)

It can be seen that in every functional respect, the magnifying glass displays the same features of the amplification-reduction structure as that displayed by the probe of the previous example. It is therefore an example of an embodiment relation in which the mediational role of the instrument occupies an embodiment position.

2. The microscope is not only a more sophisticated instrument, but begins a radical extension of the amplificational possibilities of instruments. And although in use the microscope continues to be experienced primarily as embodied vision, the amplification (and reduction) is now so enhanced that discontinuities with mundane vision also begin to appear dramatically. The micro-structure of things comes into view in such a way that much that was hidden, unsuspected, unpredicted now may be seen. But in another sense, what is seen, though appearing in a field quite transformed from the visual world of mundane perception, remains a matter of changed context and scale. (Phenomenologically, of course, this is but a change *within* the visual 'world' since what is seen is brought into the same perceptual field one sees with the naked eye. There remains at some point in each of the variations the element of mundane perception. Yet what is now within the horizon of vision, the noematic correlate, is drastically changed and what previously lay beyond the horizon has now been brought into view.)

This more radical exploration of an instrumental possibility is one which expands upon the micro-features of the World, but it also remains, in use, an expansion which experientially appears as an embodiment relation through the instrument with the micro-feature.

3. Lying behind this exploitation of a more radical possibility of the instrument is, of course, the human realization that a direction of inquiry can be undertaken through instruments. If the 'wee beasties' of drinking water were unsuspected, what can further magnification reveal? The drama of the novel within the instrumentally mediated 'world' can continue in principle indefinitely so long as humans are ingenious enough to develop more powerful instruments.

Such is part of the *telic aim* which may also be fulfilled in the use of an electron microscope. The aimed for result is still a visible result, a bringing of an even more distant micro-feature near (desevering distance). But now the means begin to change. One no longer looks through the instrument (vision no longer directly embodied *through* the instrument), instead a substitute 'eye' is used through photography. Moreover, the result is not the same; rather than the moving panorama on the slide, a *representation* (photo) is the visible result. Here, then, is a more distinct discontinuity with bodily vision. Yet the representation is still *of* the thing itself, a micro-feature of something in the World.

For purposes here it is important to note that simultaneously there begins to appear a difference in the *experienced use* of the instrument and of the *visible result*. In phenomenological terms, both noema and noesis are transformed.

With respect to the experienced use of the instrument, it may now be said to emerge from its normative, relatively transparent use, into the foreground at least during a certain stage of the investigation. At some point in the investigation I must enter into a specific relation *with* the instrument such that it is a positive feature of both my perceptual world and of my telic aim (the ultimate attainment of the visible result).

Here it may be rightly remarked that such events might also occur lower on the continuum – one first focuses the microscope, for example. But as a continuum this is a matter of degree and the act of focusing a microscope for a skilled user is very much a momentary and background phenomenon and is scarcely noted. Not so for the more complex and often tempramental electron microscope.

What begins to be obvious in this example is that the machine is functioning as mediator and this mediational functioning is becoming thematic and calls for concentration within the investigation. To be sure, this may be a 'bother' and what user does not wish for the simplicity of a directly mediated look at the object being considered? Yet the instrument now more clearly may be noted as the very condition of the possibility of gaining the knowledge sought. Its more obvious opacity is not only a matter of absence, malfunction or break-down, but is a positive element in the expansion of the visible world.

The telic result, the representation of the phenomenon in a photo, begins to produce a degree of enigma. Here the structure of transformation also attains a higher degree of discontinuity. The photo begins to lose more of what must be there (reduction) although it simultaneously does bring what was previously unseen into range.

Although the electron miscroscope is but a median example along the continuum, it brings to light what I shall call the *hermeneutic position* possible for instrumentally mediated knowledge. In terms of intentionality, the electron microscope still belongs properly to the Human-instrument-World form of relation. But the character of the mediational position of the instrument has begun to change, both with regard to its experienced use and with regard to its result. What is implicit in this change can even more obviously be noted in the following example.

4. In the case of the contemporary development of space probes, there

can be little doubt that embodiment relations are left behind and that the instrumentation itself becomes a major positive theme of the investigation. In one sense, the telic aim remains constant, say the purpose is to get a picture of the surface of Mars. In this case there will be a visible result in the representation of Mars. But I need not rehearse everything which is needed to achieve this result to point out how clearly relations *with* instruments are involved. Such relations are, in this case, so complex that whole networks of machinery, persons, and society are involved.

The launching, corrections to, and maintenance of the vehicle occupies the central concerns of the rocketry and tracking teams. Such relations with instrumentation may include embodiment relations – communications with the space probe are often exemplary of such – but the object at this stage of the investigation is itself an instrument. Then, once positioned and triggered, the means of obtaining the visible result includes a breaking up of the picture into a 'linear' transmission code, its transmission, its subsequent re-assembly into computer-enchanced series of dots, etc., which eventuate in a directly perceived photo of Mars. Both the time and the complexity of the steps taken to achieve the result are characterized by explicit relations with machines.

One must not, however, be mystified by technical complexity. All that is being illustrated here is the gradual change of what happens to the intentional arc, Human–instrument–World, in the instance of such complexity. I shall call this transformation the emergence of the *hermeneutic relation*. Hermenutic means 'to interpret' and its primary model historically is related to interpretation of *texts*. I shall use this metaphorically to elucidate what I believe to be a qualitative change in the type of mediational position occupied by the instrument at this stage of the continuum.

What the complexity of the investigative process reveals, although only different in degree from the previous example, is the obviousness and necessity of the instrumentation to be taken thematically as 'other' in the process. In order to gain a telic result, along the way relations *with* instruments become necessary if a relation *through* the complex is to be attained. Instrumentation as means thereby attains a certain phenomenological positivity which cannot be ignored. The instrumentation achieves 'a life of its own' and is a separate and distinct positive factor in the investigation.

Secondly, although the end result, the photo of Mars, displays a continuity with the previous examples of visible phenomena, it is now clearly analogous to a 'text'. The correctness of the transmission, freedom from intrusions or distortions, etc. all become part of a complex hermeneutic process.

Only in the ultimate result, after checking, correcting, and verifying, do we return to the immediacy of knowledge gained. I shall formalize this type of Human–instrument–World relation by shifting where and how the brackets occur as:

$$\text{Human} \longrightarrow (\text{instrument}-\text{world})$$

In this formalization not only is the positivity of a relation *with* the instrument as focal entity indicated, but the enigma of the hermeneutic position is identified. The symbiosis here is not so much between the observer and that which is observed, as between the veracity and reliability of the 'text', the representation produced by the instrument which is always presumably *of* the world. The instrument in this relation occupies a hermeneutic position. I must 'read' it and its result. The immediacy of embodiment relations is here displaced by the necessity of a hermeneutic process. This is not to say that such a process lacks the ease which is acquired through the skills of interpretation. It is to say that these must also be interrogated. The greater partial opacities of both the instrument and the representation which results call for the development of precisely those hermeneutic skills which are needed to 'read' the World.

B. EXPLORING INSTRUMENTAL 'INTENTIONALITIES'

To this point I have explored the gradations from immediately mediated visible phenomena (magnifying glass) towards the more mediate mediations of representations (electron microscope onwards). This does not end the series of gradations, but it has pointed up the qualitative changes which occur in a movement from the relative transparencies of embodiment relations to the relative opacities of hermeneutic relations. But in both sets of relations, the instrument retained a mediational position even though that position must be characterized differently in different cases.

However, the possibilities latent in the amplification–reduction structure are far from exhausted in the previous examples. In each of these examples to this point, a certain continuity with the mundane visible world was retained. The micro-structures or the coming into view of macro-structures which would not have appeared to the naked eye retain a certain continuity with the features of the visible world.

Instrumental mediations may also be pushed in other directions. With greater transformations, a certain *discontinuity* with mundane vision may also result. I shall try to show that with the in principle possibility of a hermeneutic relation, this greater discontinuity *frees* the instrumental use from mundane perceptual possibilities in yet other directions. In what follows, although there is in each case a visible result, it begins to be a result which is a hermeneutic *analogue* to mundane vision. It is as though textual variants were being used, but in the cases to follow there are instrumentally induced variants such that the instrumental mediation may be said to produce a different 'intentionality' from that of mundane vision.

The fifth example in the series is one which turns to deliberate variations upon representation. In infrared photography, for example satelite pictures of crop lands, by changing from straightforward representations to variants upon these (infrared), results are obtained that would not in principle be obtained by the naked eye. Mundane visibility is displaced by the emergence of an 'instrumental intentionality'.

It remains the case that the result becomes visible. Thus, as with the previous examples, what was in some sense invisible, hidden or unsuspected comes into the visible horizon, but in this case it is a feature which arises from a variation of the visual noema. I shall call the single variant discontinuity a *horizontal instrumental variant*. The photo ('text') remains continuous with the mundanely visible in that the features of the landscape are retained, the object is recognizable, etc., but a new feature has been brought into the horizon by using infrared; and we now see, for instance, where plant diseases occur on the landscape, which would not have been visible previously. A horizontal variant is an 'instrumental intentionality', i.e., is a pattern varying from mundane vision in principle, but is a single feature which is located within the continuities of the mundanely visible.

The hermeneutic quality of such representations ('texts') is now even more clearly dramatized. The hermeneutic, interpretative viewing of the observer comes into play. He must be able to 'read' the photo with regard to its significance and this reading, in turn, is informed by the 'exegetical expertise' which arises within the scientific community. The farther the continuum develops and the more extreme the variants, the more text-like the resultant reproduction becomes.

In a horizontal variation, 'instrumental intentionality' does begin to become thematic, but a horizontal variant retains visual recognizability of the thing itself. An even more extreme case, the sixth in the series, is one which transforms the representation itself through what may be metaphorically called the analytic possibility of the instrument. The example I have in mind is spectrographic photography. The visible phenomenon, a distant star visible both to the naked eye and through a telescope, may also be photographed for its spectrographic features.

Again there is a visible result, a photo which appears as a band of colored lines in a series. This visible result is obtained through instrumental mediation (of the hermeneutic sort). But strictly speaking its representational quality is transformed. The uninformed viewer would never be able to recognize the spectrographic signature as the signature of 'that star' and probably wouldn't be able to recognize it as a star at all. By this I do not mean to imply that there is some primal level of uninformed vision; to the contrary, I wish merely to point up the now strong and obvious hermeneutic quality to such instrumentally mediated features. I shall call this radical transformation of the visible in which immediately recognizable respresentation disappears, *a vertical instrumental possibility*.

The instrument itself serves as an analytic deconstructor of the phenomenom (by breaking up the light spectrum) and delivers as a 'hermeneut' its result. This result is much more 'text like' than any of the previous examples in that there is no longer any obvious correspondence of form between thing and representation. It is, rather, a 'text' which tells us something about the thing. And what it tells must now be 'read' by the one who is 'literate' in its language. What is made available, is made available through the hermeneutic use of instruments.

With both horizontal and vertical instrumental possibilities, hermeneutic relations open the way to a more radical insight into instrumentally mediated knowledge. The instrument, in these two latter examples, begins to more clearly display a distinct set of variations upon the visible, a set of variations which I have metaphorically termed 'instrumental intentionalities'. What I mean to show by this is that the deliberately designed hermeneutic functions contained in contemporary, sophisticated instruments, often deliver unexpected results, patterns and insights that, precisely because they are radical variants upon human intentionalities, take us into uncharted areas. Such uses make a more radical showing of the invisible visible. But at the same time, while in these knowledge gathering examples, the instrument in some sense retains its mediate position, it also becomes a positive phenomenon on its own. This is not to try to import a ghost into the machine, but it is to phenomenologically isolate the hermeneutic capacities of instruments as deliberately designed transformations of the mundane.

C. INSTRUMENT EMBODIED SCIENCE

It is a commonplace that one difference between classical and contemporary science is the experimental nature of contemporary science. In this context, however, it is not so much the experiment which distinguishes contemporary from classical science as its technological embodiment. Whereas classical science was limited for the most part to speculation, theory, deductive cleverness and primitive measurements, none of which are absent from contemporary science, the technological instrumentation now available allows inquiry to be extended in ways never dreamed by the ancients. Moreover, what shows itself even with the limits of the continuum explored here, is the indefinite extension of probing the macro and micro features of the world made possible through instrumentation.

It is through instrumental possibilities that the vision of science is extended, both in terms of directly mediated embodiment and the mediated mediation of hermeneutics. But if these modes of investigation are the condition of the possibility of scientific inquiry, are they also more particularly formative of that inquiry? The possible, latent telos which lies in the technology of instrumentation will be taken up in chapter 4.

There remains, however, a preliminary set of suggestions which sets the stage for a second level inquiry into telic aims implicit in the technics of instrumentation. The qualitative differences between embodiment and hermeneutic relations also suggest two quite different trajectories for or investigation. These may be called a trajectory toward the perceptual and a trajectory away from the perceptual. Both are present in instrument-embodied science.

The limits I have set in this essay remain those which culminate in some form of visibility, however transformed. Furthermore, some element of direct perception remains operative even at its most hermeneutic extreme, necessary within the 'reading' process. But even within these limits the shift from directly mediated embodiment relations to indirectly mediated hermeneutic relations has been indicated. The formalizations which characterize embodiment and hermeneutic positions for the instrument are indicative:

Embodiment Relations: (Human—instrument) ⟶ World

Hermeneutic Relations: Human ⟶ (instrument—World)

Intentionality is transformed in the shift from one to the other. And in use the phenomenological appearance of the instrument also changes as does, gradually, the fulfillment of any intentional aim, the object correlate. Thus, again for every noematic change there is a corresponding noetic change.

The two trajectories I have suggested may be indicated in terms of what first appeared as an ideality on the embodiment side, the idea of pure transparency, in contrast with the emergent notion of an 'instrumental intentionality' at the hermeneutic extreme. In the first case the ideality is approached if and when the instrument phenomenologically 'withdraws' or attains transparency in its being used. But in the appearance of an 'instrumental intentionality' the instrument must necessarily appear as a quasi-other with some degree and stage of relative opacity.

Thus what becomes epistemologically enigmatic also shifts in the two types of relations. In the case of embodiment relations the symbiosis (the dash between Human–instrument) must be carefully analyzed lest one forget the transformations of direct perception which occur. On the hermeneutic side, the nature of the connection between the instrument and its object (the dash between instrument–World) may become extremely enigmatic particularly because there is an *unexperienced* opacity here. For example, while the versimilitude apparent even in a horizontal possibility leads one to take the representation as genuinely *of* the object, the transformation involved in the 'instrumental intentionality' remains enigmatic. What now characterizes the amplification structure? And what is forgotten in its reductive convexity? In the simple example of infrared photography there always remains the dialectic of what is seen with the naked eye and what is produced through the instrumental variation. But with vertical instrumental possibilities even the versimilitude disappears and the instrumental opacity becomes more fully hermeneutic.

It remains the case that in either direction, towards or away from perception, the amplification–reduction structure is invariant, but its shape changes from case to case. At the embodiment end of the continuum there is retained a similarity between mundane perception and what is brought into the visual horizon. The micro-organisms of the microscope slide, though never before seen, appear as motile, colored shaped entities even though the focal plane of the microscope reduces their three-dimensionality and changes in subtle degrees the visual environment in which they are seen. I shall characterize this shape of amplification-reduction as one of *low contrast*.

On the hermeneutic side of the continuum, however, there appears a *high contrast* amplification–reduction shape. In the case of the spectrographic representation the 'analytic intentionality' of the instrument reduces the visible to what may be called a mono-dimension. This mono-dimension is highly amplified in the instrumental transformation – that is precisely what makes it valuable for knowledge gathering – but at the same time the reduction is equally dramatic in that the 'object' in an ordinary sense disappears so far as recognizability is concerned. What remains is the instrumentally delivered 'text' which is now 'read' by the scientist—hermeneut.

In the case of both horizontal and vertical instrumental possibilities the high contrast amplification—reduction structure is tied both to instrumental opacity and to the emergence of 'instrumental intentionality'. The deliberately designed transformation from mundane perception is the condition of the possibility for the emergence of certain analytic functions through the instrument.

I have suggested that two quite different trajectories may be associated with the two extremes of the continuum. Both of these directions may also be detected within the metaphysics and praxis of contemporary science. Often the two directions are co-contemporary without a sense of divergence or contradiction. The hermeneutic extreme responds to a long and deeply held metaphysics which originates in idea with Democritus. The division of sense (perception) and reason (hypothetical deduction) posits an ultimate, unsensed unexperienced 'world' as ultimate; and its existence is not merely that of in practice unperceivability, but of in principle unperceivability. Yet in contemporary technologically embodied science, it is precisely what is thought to be unperceivable that is made *present*. The successful search for atomic subparticles, for DNA structures, for the various varieties of the micro-structure of the world *appear* through the mediation of the sophisticated instrument.

But their appearance in the hermeneutic relation is both indirect and highly transformed. The emergence of 'instrumental intentionalities' which transform what is made visible, the appearance of high contrast amplification—reduction shapes, and the analytic use of instruments which reduce to monodimensions, are often the only ways to gather knowledge about the highly micro or macro elements of the World.

Yet through all of this there is a certain amount of uneasiness for having seemingly left the realm of mundane perception. Democritus, too, felt the unease. He says of the senses.

Ah, wretched intellect, you get your evidence only as we give it to you, and yet you try to overthrow us. That overthrow will be your downfall.[1]

Thus simultaneously within the technology of instrumentation, there remains the desire to move in the direction toward transparency with the secret wish and belief that somehow to 'see the things themselves' is the ultimate and genuine form of knowledge. After and perhaps beyond the hermeneutic inquiry lies the hope that what is discovered can ultimately be seen.

I may illustrate this direction in terms of a perceptual trajectory of instrumental development, a development which seeks to maximize the ideal of the perfectly transparent instrument. This direction is most obvious when there is any bodily involvement with what is being worked with. For example, in the development of equipment constructed to work with radioactive materials at a distance (mechanical arms), it becomes desirable to perfect them so that 'feedback' is such that the operator can feel through the machine the delicate resistance of the glass container being moved by the robot arms. Or, in the case of developing navigational equipment for aircraft, the desirable development of hermeneutically positioned instruments (dials and gauges) is considered better if a visual representation of the runway (radar-informed television) can appear before the pilot.

But the same tendency occurs in scientific investigations. The recent developments of 3–D photography (holography) and of computer–enchanced amplifications which supposedly even produce pictures of molecules and now even an atom illustrate the urge to 'see the thing itself'. The latent telos of the return to perception is one which seeks to enhance the quasi–transparency of instrumental relations toward the range of mundane perception.

Yet both directions are explorations made possible through the development and embodiment of knowledge-gathering through instruments. Contemporary science is technologically embodied science. Some, most notably Heidegger, would make a more radical claim: contemporary science is technologically formed. That is a question which calls for an investigation into the telos of contemporary technics.

NOTE

[1] Philip Wheelwright, The Presocratics (New York: Odyssy Press, 1966) p. 182.

CHAPTER 4

A PHENOMENOLOGY OF INSTRUMENTATION:

Technics and Telos

That the instrument is an element within contemporary scientific inquiry and that it occupies some type of crucial mediational position upon the intentional arc has now been shown. But if contemporary science is embodied through its technics a question can arise as to whether instruments play any deeper, *reflexive* role for the formation and development of subsequent inquiry. The question itself can be put simply: once having begun an inquiry through and with instruments, does the range of instrumental capacities incline the inquiry in certain rather than other directions?

This is necessarily a much more speculative question than those which motivated the previous two parts of this phenomenology and thus some concern over strategy must be included here. The focus of a strictly phenomenological analysis, in this instance, remains upon the instrument within its experiential context. Such a focus does have the advantage of quickly demythologizing the types of naïvete which would confuse pure or ideal possibilities with their actually experienced base. Thus the 'subjectification' of the instrument which would see it as a pure transparency such that the instrument is regarded as a mere neutral tool should have by now been thoroughly demythologized. But at this juncture a second danger emerges at the opposite side of the issue which threatens an equal confusion. It is also possible to 'objectify' the instrument such that it is understood to be animated, to have its own 'ghost', and thus be reified. There are today many positions regarding technology which fall precisely into this form of mystification, which see in technics the emergence of a Frankenstein phenomenon which outstrips its creator and eventually turns upon its creator destructively.

These issues, which are implicit in much of the general discussion of technology, take more precise form in the examination of instrumentation. By concentrating upon this narrower set of artifacts in use it is possible to isolate more exactly where and how these confusions might arise. I have noted that there are two counterpart confusions which are to be avoided.

Both of these have their roots in a pre-phenomenological understanding which confuses what occurs within the limits of intentionality. The 'subjectifying' confusion which takes its dream from 'pure transparency' leads to a utopian interpretation of technology. This confusion overlooks the reductive dimension of the instrumental transformation structure and dwells only upon the more obvious drama of transparency. Thus the 'semi' of semi-transparency is overlooked and confused with the dream of ideality which is itself never present in intentionality. This partiality of the instrument's transparency is detected only through the critical application of the analysis.

To return to the paradigm example of the dentist's probe, what is focal and foreground is what is experienced *through* the probe: the tooth's surface. Yet even while this occurs it is possible to take note of the 'echo focus' which points up the presence of the tool itself in use. I always feel the slight pressure of the tool upon my fingers co-present with the more dramatic presence of tooth's surface which is the ordinary intentional terminus of the experience. The tool never completely disappears and only the *dream* or the ideal of pure transparency occurs within bodily experience.

I shall not here unpack what may be the root causes of the confusion between the dream of pure transparency and the phenomenologically isolated actuality of partial instrumental transparency, but one might be led to expect that the philosopher's secret desire for a purely disembodied state lurks behind such confusions.

Such is the general shape of the confusion which arises with the 'subjectifying' tendency which reduces the instrument to pure neutrality and thus allows for the utopian possibilities that arise from such positions. There is, however, a symmetrical counter-confusion which occurs at the opposite end of the continuum of instrumental possibilities. An instrument in hermeneutic position phenomenologically emerges from its transparency and takes on a positivity which makes the instrument a *quasi-other*. Thus, at least during those parts of the investigation in which I engage the instrument, it appears as a type of Other in the process. With the emergence of horizontal and vertical 'instrumental intentionalities' the temptation for reification increases. The very opacity of the unexperienced instrument—world relation places a source for mystification before us. As the drama of amplification unfolds at the hermeneutic end of the instrumental continuum, the temptation to forget the hidden reductivities implicit in instrumental use arises. Here the confusion is an exact counterpart of the 'subjectifying' confusion in that the 'quasi' of quasi-otherness is forgotten. What is needed is a critical reflexivity which interrogates the process of 'reading' so that instrumental transformations are noted to have two dimensions rather than one.

Phenomenologically, what has been isolated with respect to both embodiment and hermeneutic instrumental positions is the invariance of the transformational structure. The instrument, in use, is non-neutral and this non-neutrality takes specific shape through the inquiry. It is by interrogating further this non-neutrality that I hope to point up the *latent telos* of technics. There are three principal steps to be taken: (a) in keeping with the strategy of concrete phenomenological descriptions, I shall first examine examples of latent telos in technics. Both embodiment and hermeneutic examples will be elucidated. (b) I shall then turn to a more general projection from the already demonstrated structural features of instrumental transformations and try to show how a general latent telos may effect the formation and development of scientific inquiry.

A. LATENT TELICS IN INSTRUMENTS

Although some positions regarding technics take what might be called a hard 'technological determinist' line, I do not believe the phenomenological evidence warrants such a conclusion. A hard technological determinism would have to demonstrate that the use of given instruments or sets of instruments so determines an inquiry that only certain directions are possible rather than others. I do not believe that the evidence provided by a critical analysis leads to a hard deterministic conclusion, but what it does yield is evidence that there are latent telic *inclinations* which are made possible through the use of instruments, inclinations which favor certain rather than other directions. I shall take two examples, one from the use of instruments in embodiment position, the other from a hermeneutic position, to illustrate such a phenomenon. The examples chosen are not, strictly speaking, examples of a scientific inquiry in action, but are more ordinary experiences which do show the same phenomenon.

To compose, in this case to write an article, some instrument is used which on the occasion of composing is usually positioned within an embodiment relation. Thus whether I compose by pen or by typewriter the instrument recedes and displays its nonthematic and semitransparent appearance. The terminus of the experience, that which is thematic, is the work, the appearance upon the paper of that which is being expressed. Thus with both pen and typewriter the embodiment symbolism is maintained:

(Human—instrument) ⟶ World

And, in both cases, the invariant features of the general transformation obtain. But the particular shape of that transformation is quite different in the two cases, or, as I will say here, each instrument presents a different set of possible inclinations for utilization.

A most obvious factor to note in this illustration is the difference in speed of writing which is made possible by the two different instruments. The person skilled at composition by typewriter can very rapidly place his thoughts on paper, while the user of pens (unless a shorthand expert) necessarily lags behind in the amount written with the same time period.

The relative speeds of composition, however, can effect the *style* of writing. For example, the composer using the pen has time, before finishing a sentence, to think through a series of possibilities and make choices among these before they ever appear as a completed sentence. The rhythm of the pen is slow and enhances the deliberation time which goes into writing. Contrarily, the typewriter composer, if the rhythm of the instrument is to be maintained, finds that almost as soon as the thought occurs it appears upon the papers. To make a bold contrast, the telic possibilities of the instrument which may incline the user to a certain style, are those which favor something like *belles lettres* in the use of a pen and a more colloquial or journalistic style in the use of the typewriter.

Now it is equally obvious that such a telic inclination which is made possible by the different capacities of the instruments is *not* a hard determinism. The user of the pen can produce a colloquial, journalistic style just as the user of the typewriter can produce the deliberate effect of *belles lettres*. But, the kind of effort which is demanded for these results is noticably different and often obtained differently. The telic inclination made possible by the instrument does not cut off any human aim through itself, although it does call for varying degrees of effort on the part of the user to counter whatever may be the implicit rhythm of the instrument in its normative and functionally optimal use.

What this example purports to show is that over time, over practiced use and in general, the telic inclination made possible by the instrument creates a path of least resistance or of highest functionality which *may* be followed and often *is* followed. The instrument provides the condition of the possibility of an *instrumental style* through its latent telic inclination.

A counterpart phenomenon occurs within the hermeneutic relation. Here the telic inclination is one which finds its fulfillment in what I shall call the *realism* of instrumental results. The temptation to succumb to what may be characterized as the ease of taking the path of least resistance is the most obvious hermeneutic result.

My example here is one which is hermeneutic in that the reproduction (text) which is produced through the instrument purports to be a reproduction of what is independently experienced without the mediation of the instrument. I have in mind the use of a tape recorder. If the tape recorder is an ordinary one with a wide range or omnidirectional microphone and it is turned on to record a speech or paper being given in a large auditorium at a philosophy convention, unless the mike is placed close to the speaker the following result occurs. The speech, which was easily heard by the ordinary listener, may on the tape recorder be barely heard at all. Instead, what were for the listener barely noticed background noises such as the shuffling of feet, coughs, the hum of the air conditioner, etc., come through on the tape as obviously as strong as, or even overwhelm, what comes through as the speech.

This is to say, using the analogy developed in the hermeneutic examples, that the instrument has a quite different 'intentionality' than that of the human listener. The 'world' of the tape recorder is almost that of a sense–datum empiricist where all external stimuli are reduced to an equivalence and thus none are more meaningful than others − only in this case the human listener to the tape can no longer recover what he heard quite effortlessly in the same situation. Here an opacity between instrument and its referent has occured such that even a 'reading' of the 'text' becomes difficult if not impossible.

There are many directions which this example may lead, but the direction I wish to point up is somewhat curious in that the 'world' which the tape recorder presents *may* be taken as *the real*. Certainly the sounds it recorded were there (shufflings, coughs, etc.), but they are, compared to direct hearing, grossly transformed by the instrument. Yet the possibility of *taking* the 'world' presented by the tape recorder as real is raised and is often followed as such in scientific investigations.

The realism which is made possible here is one which reads the hermeneutic text 'literally', as directly representational, rather than critically, as a transformed representation. Such a *style* of reading is made possible by the telic inclination of the instrument.

In both the cases examined it is apparent that instruments in use present their users with subtle, often undetected and yet present possibilities of inclination towards one rather than another style of inquiry. But at the same time, it is also apparent that counter measures, efforts, could be made to resist or overcome any such inclinations provided by the instruments. Contrarily, for the most functional and optimal use of the instrument it becomes likely that in practice the telic inclination will be followed with a subsequent change in style depending upon what the instrumental complex may be. It is from this base that the larger question of a latent telics of instrumentation may arise.

B. PROJECTING TELICS

What follows now is a more hazardous step from the previous limited examples to a speculative generalization about overall telic inclinations within the domain of scientific instrumentation. Obviously there is need for caution here in that the previous examples have already demonstrated that different instruments display different telic possibilities. Thus this speculation is necessarily less rigorous than a straight forward phenomenological analysis, but at the same time I think that the suggestions which follow regarding the embodiment of contemporary science could be shown.

I shall put the speculation in an imagined, hypothetical setting: what if one regarded the previously described invariants of instrumental uses as directly pointed towards a telic aim? What would a 'world' look like if it were instrumentally constituted? I shall try to project such a constitution of a 'world' from the aspects of the transformational features previously noted.

The overall shape of the instrumental transformational structure is that of an invariant amplification–reduction. It was noted that the amplifacatory dimension appeared as dramatic, the condition both for novel features of the world to be made present within the human experiential horizon, and the source of fascination which might serve as a motivation for further investigation. Even in the grossest and simplest way, it is possible to see here a latent telic possibility. In the visual examples, if some magnification yields the previously unseen, will not greater magnification yield even greater wonders? Here there is a latent direction presented for the development and perfection of an inquiry.

Speculatively, it may not even be accidental that a visual world became central. It has already been noted that the technics of early modern science correspond with the then new optical technologies which provided telescope and microscope as primary scientific tools.

While it is easy to understand the fascination and drama which accompanies the amplificatory dimension of instrumental technics, phenomenologically it remains the case that the other dimension of instrumental transformation is reductive. But, as previously noted, instrumental reductivity is frequently a background aspect which may be 'forgotten'. The features of noema which 'drop out' in the instrumental investigations may be ignored. But what if the forgetfulness of the reductive effect were to be made systematic? What then happens to the 'world' which is being instrumentally constituted?

In the hypothethical situation being imagined here, assume that the technological successes are limited to the invention, development and use of optical instruments and their variants. Were that to be the case the beginning emergent 'world' would increasingly become a visual 'world'. The other sensory dimensions would gradually take a position graded lower and lower on the scale of 'reality yielding' capacities and, correspondingly, the visible features of the world would be taken to be more and more highly valued.

It might seem that such a state of affairs would be unlikely. The reason being that in the mundane perceptual world it remains the case that I go about not only seeing things, but handling, touching, tasting, hearing and smelling them. And, as long as the access to the various things in the world remains roughly symmetrical with what I see through the optical instruments, I might retain my overall sense of the 'sameness' of the mundane and the instrumentally mediated world.

But what if the technics of visual instruments, stimulated by my fascination with the novel features I find through them, begins to develop a *high contrast* amplification—reduction such as those previously noted, so that both the easy recognition of a mundane world recedes and the analytic capacity of the instrument begins to become apparent in the reduction of what is seen to a monodimension? Here the discontinuity between what I mundanely experience and what I see through the instrument begins to be heightened so much that a temptation arises to posit two 'worlds', one a mundane world, and another an instrumentally constituted 'world' which is only accessible through or by means of the instrument.

In this hypothetical construction, I am now faced with a serious discontinuity between the full sensory world of mundane experience and the now monosensory 'world' of what is instrumentally mediated. Suppose, now, that I begin to believe that in the difference, one of these worlds is more real than the other. Suppose that I believe that the instrument is the very model of precision and perfection. Then I begin to believe that my eyesight is very limited, that it hides more from me than I see. And the superiority of the instrument mediated 'world' is what lies underneath (if not behind) the mundane world. In short, I begin to accept, literally, the instrument mediated 'world' through what may be called here an *instrumental realism*. The instrumentally constituted 'world' becomes the 'real' world. Not only do I forget the mundane world, but it begins to be downgraded.

But I am now faced with a problem: the relatively effortless aquaintance I have with things in the mundance world 'phenomenally' reveal aspects to me which are present through sensory dimensions other than visual and my instruments are all optical. It may still be the case that I can adumbrate my instrumentally constituted 'world' to account for everything I find in the mundane world. To accomplish this I invent and develop an instrumentation which not only reduces the world to a visual 'world', but an instrumentation which 'translates' all other aspects of the world into visible results. Thus if I investigate auditory phenomena, I make them visible through a vertical instrumental possibility and a voice pattern becomes a visible pattern on an oscilliscope, etc. I am now on my way towards the construction of a totally mono-dimensioned 'world' which through my belief in an instrumental realism, is taken for the 'real' world. I am here suggesting that the latent telos of such an example may enhance a certain type of metaphysics. Yet, ironically, it is enhanced precisely because the instrument has a hidden 'phenomenological' capacity. Instrumental realism becomes possible and even convincing *because* what was invisible becomes *present*, becomes a fulfillable noema.

This hypothetical constitution of an instrumentally mediated 'world' does not *demonstrate* the technological ground for contemporary scientific investigation, although it may suggest a closer correlation of technics and science than is ordinarily thought to be the case. A more convincing speculation arises from yet another feature of the invariant transformational structure of instrument use.

It was noted that instrumental mediation transforms the shape and distance of the world. The desevering or bringing into presence of that which was previously either unnoticed or undetected is done in such a way that micro—macro features of the world are made focal. The instrument, metaphorically, concentrates upon the micro—macro features of the world. This structural component of instrumental transformation was, of course, noted as latent in the previous characterization. But here this feature taken as a telic possibility of instrumentally embodied investigations becomes more obvious as an actual feature of contemporary science.

That contemporary scientific investigations are highly committed to the examination of extreme micro-features is abundantly apparent. Particle theory in physics, DNA and genetic investigations in biology, and the construction of elements from micro-phenomena in chemistry are all familiar. What is not noted so strongly is that these areas of investigation are frequently regarded as the highest prestige areas of frontier research in the sciences. Such investigations are, of course, not possible without increasingly sophisticated and — perhaps not incidentally — *large* instruments. My suggestion is that not only is the condition of the possibility of such investigations related to technics, but that the latent telos of such technics provides in part the ground for such investigations.

Nor are the features suggested in the previously more hypothetical example absent. I am told by psychologist friends that the same picture obtains in psychology. Not only are investigations and the prestige associated with them largely those concerned with micro-processes (neurological psychology is the most obvious, but even studies in perception are now largely concerned with micro-processes). Moreover, such studies in psychology are also technologically embodied. To examine a subject's response to a stimulus, one employs timing devices which break down the temporal span into micro-components, measuring, for example, the minute differences of time that it takes a subject to recognize a configuration containing four rather than three dots. Perhaps the most direct and evident example of scientific investigation correlated with instrumental telics is this concentration upon micro-phenomena.

Here, too, the features of the implicit belief in instrumental realism occur within the social dimension of the scientific community. The belief that the 'world' mediated through the instrument is more 'real' in its analytic and micro-dimension pervades much of the theoretical concern illustrated in the aforementioned examples.

The drama of instrumental amplification continues to permeate instrumentally embodied inquiry and offers the temptation for the forgetfulness which overlooks reductivity. Does this concentration as a characterization of the dominant and frontier research lead to 'forgetfulness' of what might be equally interesting, important and promising other dimensions of the world? There are many who express this unease even within the scientific community.

In this hypothetical pairing of an instrument-constituted 'world' with its latent telos for investigative directions I do not claim to have phenomenologically grounded the current form of science. Nor have I demonstrated that Heidegger's more radical claim that technology is more originary than theoretical science is the case. I have suggested, however, that there may be more involved in the use of instrumentation than a matter of mere application of theory to practice.

By projecting a state of affairs from the non-neutrality structure of instrumental transformations, I have suggested that a straight forward social determinism such that instruments might be thought to embody just any human aim or interest is not adequate. Instruments embody human aims and interests in certain ways, ways in keeping with the necessarily transformational characteristics of the amplification—reduction structure. On the other side, no technological determinism in the hard sense is adequate either, since technics in its telic dimension provides only a base for inclination rather than determination in any hard sense. Neither of these parameters, however, preclude what may appear as a center of gravity which allows a direction to be followed from the inclined possibility structure of technics.

C. EPILOGUE: PHENOMENOLOGICAL REALISM

In this phenomenology of instrumentation, the instrument has displayed itself as simultaneously the condition of the possibility of certain types of knowledge and yet as a non-neutral transformation of what is known. In use the instrument has the 'phenomenological' capacity to bring into presence that which was previously undetected and even invisible, but precisely in this *difference* it also transforms the way in which the phenomenon may appear.

Yet within this enigma there may be detected a type of 'realism' in that technics makes the world visible. Technics is a *variant* seeing, but variants within phenomenology are variants upon the phenomenon. It was necessary to point up the enigma of technologically embodied investigations in order to guard against the temptations to 'subjectify' the instrument.

There is no pure transparency to be found. Equally, the recognition of the transformational structure of instrumentally mediated knowledge cuts off the opposite tendency to idealize the instrument in a myth of perfection. Critical awareness of the forgetfulness which is possible concerning instrumental reduction must be maintained.

Still, the result is one which continues to underline the first 'phenomenological' capacity of an instrumentally embodied investigation to get to the 'things themselves'. There is also a second 'phenomenological' capacity which is possible in technics. Once forewarned against the effects of transformation, an *informed* investigation can enhance the 'phenomenological' capacity of instruments by developing ever richer sets of instrumental *variations*.

By such adumbrations the implicit reductive dimension of instrumental transformations can be partially corrected. Note a very frequent pattern to be detected in contemporary investigations of phenomena previously observed through very limited instrumental means. Until very recent times observations of Mars and the Moon were restricted to strictly visual instruments (including, of course, those of high contrast structures and those of extreme hermeneutic use such as spectrographs). In the case of the space probe sent to Mars the actual penetration of the soil (tactile extension in hermeneutic position) with the subsequent chemical analysis (with analogues to tactile, olifactory and even gustatory capacities) yielded some highly surprising effects still not fully understood.

There are two aspects to this surprise. One aspect is quite ordinary: we could not expect to know exactly what the essence of the Martian surface was like until we actually 'got to it'. But the other aspect is one which is so often repeated that it may reflect something of the depth effect of our expectations guided by instrumental telics. Whereas we thought we pretty well knew what to expect because we had had a very sophisticated (visual) contact with Mars, it turned out to be more complex than we expected once it was brought more fully into our presence. In short, I am suggesting that just as our global sensory experience exceeds any reduction to mere visual experience, so an adumbrated investigation exceeds the necessarily reductive effects of especially high contrast, mono-dimensioned instrumental investigations. A similar survey of the impact of 'auditory' extensions in the form of sonar, radio astronomy, and the like which now make present such phenomena as temperature inversions, whale communications, and black holes, may be considered here as a technologically embodied 'phenomenological' variation.

Behind these comments may be detected an implicit paradigm in which a form of the 'primacy of perception' may be seen to lurk. To that I confess, but such a paradigm also has a heuristic aim which is instructive in the various examples examined in this phenomenology of instrumentation. If I suspect in advance that the world is rich, complex and open to the ultimate perceptions of human experience, then I can also seek ways in which those perceptions are embodied such that the various reductions and temptations to reify partial aspects into totalities can be critically guarded against.

DIVISION TWO

IMPLICATIONS OF TECHNOLOGY

CHAPTER 5

THE EXISTENTIAL IMPORT OF COMPUTER TECHNOLOGY

The task of this essay is to reflect upon the non-technical experience of computer technology. In what is to follow I shall elaborate upon a number of theses which, in a general sense, apply to *any* technology, but in this context will be directed more specifically upon computers and their experienced results. The theses are: (1) *Any use of technology is non-neutral*. The term, non-neutral, is deliberately chosen to indicate that the use of technological artifacts transform experience in some way. I do not wish to imply either negative or positive values for the transformation as such, but wish to underline that there is a significant transformation of experience in the use of technologies. (2) Within overall experience, there are a number of primary categories such that the user of a technology may experience technological artifacts in several different ways depending upon how the artifact is related to the user. (3) Subsumed under the first two theses there may be discerned certain specific characteristics of the transformation of experience. Technologies organize, select and focus the environment through various transformational structures to be outlined. (4) From the variety and structure of transformed experience, the existential import of computer technology may be glimpsed. It takes the shape of what I shall call a *world-reflexivity*, a term which will be clarified en route.

(1) *The use of technology is non-neutral, it transforms experience.* If I take as my normative paradigm of experience an unmediated perception of the environment, then what is directly touched, grasped, tasted, seen, heard, forms the surrounding world of direct or 'fleshly' experience. The apple looks shiny red, smells 'applish', tastes sweet and crunchy, and even sounds firm to the tap. My primitive contact with nature, others, the environment, is often direct — but it is not always so. Many of my perceptions are transformed in a mediation, a mediation which in some way employs, encounters, or engages some form of material technological artifact.

If the bunch of bananas in my elicited primitive 'paradise' is beyond my **bodily** reach, I may employ a stick — or even better, a stick with a knife attached — to get them down so that I may eat them. The stick *mediates* my relation with the bananas in the moment I use it and at the moment of cutting down the bunch. My experience of the bananas is markedly (though probably not noticed prereflectively) different than what I would have experienced had I done the task by hand.

In a few moments I shall examine some samples of what this transformation entails, but note that even in this simple example the stick with a knife 'extends' the reach of my body; transforms the perceptual experience of 'getting bananas'; and introduces a certain factor of selectivity into the action (for instance, relative ease or lack of difficulty with the ensuing value of 'efficiency').

What the illustration shows is that if there was a 'technological fall' it occurred at a very primitive stage of humanity; exists among lower animals; and has for practical purposes always been among us in principle. But on the other hand in a technologically saturated society the sheer number and type of such human—machine uses has expanded exponentially to the point that it becomes very difficult to even ennumerate the number of such uses in a given day.

However, the point here is both a simple and general one: the very use of a technological artifact, be it stick with knife or computer, is non-neutral, it in some way transforms direct or non-mediated experience in ways yet to be determined.

(2) *Types of experience with technologies*. While I shall maintain that *any* use of technology is non-neutral, a first set of distinctions needs to be made with repsect to a set of possible types of experiences of technology.

There is first what may be called experience *through* a technology. In this case I mean the type of experiences in which some artifact is used in such a way that something else is experienced *through* the artifact itself. The previous example of the stick with a knife is one such experience — I experienced the bananas through the stick, I felt the stem being cut, and I can even say descriptively that I felt the stem being cut *at the end of the stick itself* at the very juncture of knife—stem. In short, the artifact in this case extended my self or *bodily* self experience through it and I became 'embodied' at a distance and experienced this genuinely, although **mediatedly**.

Such experiences of the world through technologies are often expressively exciting. Take the sense of discovery which we are all familiar with in our first learning how to drive, the first look at the mountains of the moon or the rings of Saturn through a telescope, etc. While there are important invariant structures to this experience, all that is being noted preliminarily here is that this experience of technology is one of experiencing something else *through* the technology being used, and that *what* is experienced through the technology is a part of *perceptual*, bodily experience.

There is another type of experience, however, which is markedly different. It is what I shall call the experience *with* a technological artifact. If the first case is one which uses a technology as a *means* of experience, the second case is one in which the technology is itself that which is experienced, although in any use I do not want to say that it is merely an 'object' of experience. Here my favorite example is one which my daughter experienced some years ago: when she was nine, she joined a research project which was studying some early teaching machines composed of a programmed set of mathematics examples hooked into a computer run typewriter. She would turn on the machine which, in turn, printed out a problem. She would then have to respond and attempt to solve the problem. The program was such that at each step, particularly difficult ones, the machine would type a 'reenforcing' compliment ("You have done very well, go on to the next step"). But also, when the user became tired and began to make errors, after a certain number of these the machine would type out, "You seem to be tired now, why don't you go home and come back tomorrow to finish." Needless to say, my daughter was both intrigued and amused by this human—machine relation.

But what is important to note in this case is that the relation is one which is a relation *with* a machine. In fact, there is something of a very primitive 'dialogue' going on here in which the human user 'reads' from the machine and responds to what is read. The machine, rather than being a partial extension of myself, my bodily self, is in this case a quasi-other. The teaching machine appears as a kind of — rigid to be sure — other with a highly technical, but limited vocabulary.

In this particular context it is preliminarily important to note that a good amount of the *technical* experience of computer technology is of this second type. The programmer, the user in a day-to-day fashion, may relate to the machine as a 'something' or even a quasi- 'someone' to whom to relate.

Both of the above types of experience of technology are *focal* experiences in the sense that either one is relating directly *through* a machine to something in the world, or one is relating *to* a machine as something directly within my attention within the world. Yet a third type of the experience of technology is one which is neither focal nor so direct, it is what I shall call a background relation.

An example of such relations is not far from us. According to modern architects, we live and work within a 'machine for living'. And this artifact is filled with other artifacts which provide a background within which our focal activities may take place. The lighting, the heating, the acoustical engineering with all its semi and usually automatic activities form the immediate background, the 'technological texture' to our daily activities. Unless they malfunction, we seldom note this background explicitly. And, indeed, the machinery which makes them run is even hidden from us, probably rather far from the immediacy of the here and now.

Background relations are *field* relations, they are the source of the general texture of life and in the example given are the condition of the possibility of our comfort and communication. But in this particular assignment concerning computers, I think it will be apparent that the computer technology which is somehow both 'behind the scene' and yet also 'active', is what constitutes much of the experience of the *non-technical experience* of computer technology. And the point made above concerning the ordinary familiarity and forgetfulness which characterizes the way we take background relations for granted *until they fail or obtrude* applies with particular importance here.

To this juncture I have developed a very general framework concerning the experience of technologies to indicate both that experience is transformed in the use of technologies and that the varieties of experience are in fact coextensive with what may be called the range from our self experience (experience *through* machines), of that which is other (experience *with* machines) and of the general background of the immediate environment or world (what I shall call experience *among* machines). What next needs notice are some specific features of those transformations and their structures if we are to begin to note the existential import of computer technology.

(3) *Structures of transformation, the specific shapes of non-neutrality*: The primary structure of transformation to which I wish to draw attention is what I shall call the *amplification–reduction–*transformation. By this I mean that the experienced used of technologies brings with it a simultaneous amplification of certain possibilities of experience while at the same time reducing others. Not only do these effects occur simultaneously, but they belong necessarily together. Non-neutrality consists in part in this transformational structure.

I may illustrate this simply from the very first example (the stick with knife) by recalling that the stick 'amplified' certain bodily possibilities, it 'extended' my reach. This aspect of the use of machines is dramatic and seldom overlooked, it is obvious and one of the reasons why ever new technologies seem so appealing. Amplification stands out.

But simultaneously and necessarily connected, I have indicated there is a reduction of experience. In the case of the stick-knife, I cease to experience the toughness and sinew toughness of the stem, it recedes from my direct experience and with it part of the sense of the 'living banana'. But what I am 'forgetting' or 'missing' can easily occur so that I do not notice the reduction until something else occurs to recall it to mind. Yet in reality, the full banana-experience potentially contains that which is being reduced.

Secondly, this simultaneous amplification—reduction forms the peculiar 'selectivity' of the technology. Its effects with respect to experience-possibilities may be illustrated in a simple example: When I was in France a number of years ago, my children were enrolled in the French public schools and the mode of teaching writing was through the use of the old dip pen. In the evenings I began to play with those pens through which there seemed to flow a visually delightful line. But also accustomed as many of us are to composing on the typewriter, I discovered that the ease and speed of writing was slowed dramatically. With the typewriter, thoughts flowed rather unimpededly through to the paper, but with the pen one had the time to reformulate a sentence many times before ever reaching the end. In the process I began to discover a difference between the short, clipped sentences of speed typing and the inclination towards the old style of *belle lettres* through the pen. This is not to say the technology *determined* the style, but it certainly *inclined* it. Its different ratio of amplification to reduction in the two instruments displayed selectivities within which experience could take shape. And in a sense, only a disciplined attempt to resist the inclination could keep the sedimentation of an inclination from forming.

This general observation concerning amplification and reduction and a shape of selectivity applies to computer technology as well. First, note a very broad selectivity concerning computers: out of the entire possible range of human experience, the development of computers *selects* a certain range. Furthermore, this selection, in part deliberate on the part of its designers, but also in part appropriate to the artifact itself, provides the basis for a much farther reaching *inclination* for the possibilities of human social and existential experience.

This amplification—reduction—selectivity may be made to stand out if one places on a continuum certain aspects of human experience. On one end of the scale I place a set of human capacities and experiences which are often *unspoken*, for example, bodily finesse, the kind of graceful 'know-how' developed by a dancer or even the ordinary skiier. I include such things as very discriminate 'gestalt judgments', the kinds of takings of wholes which we do daily in matters of taste and discrimination. I include ordinary types of pattern recognition, such as picking out of a whole crowd someone I knew long ago and who may even look like someone else. On the other end of the continuum I place activities such as counting, adding sums or more complex calculations, the development of tightly knit logic trees, the making of deductions, and even the playing of simple finite sum games.

I am not here interested in entering the thickets of the artificial intelligence debate, nor am I interested at this point in arguing ultimate possibilities or impossibilities of modelling along this continuum, but persons as diverse as Hubert Dreyfus, the arch critic of artificial intelligence, and John Truxall, Stony Brook's most vociferous defender of technology, agree that the computer does very well at one end of this continuum and does poorly and primitively at the other. My only point is a simple one, computer technology in its capacities is both selective out of the range of analogues of human experience possibilities, and is amplificatory and reductive within that selection.

This is, however, a general point concerning any technology. What is needed here is a more specific look at the kinds of amplification–reduction which comes from *interaction* with the computer. First, in terms of the outline of the categories of experience related to machines, for the most part the *technical* experience of the computer falls within the second category, experience *with* a machine. The user, the programmer, the reader relate in various ways actively with the computer so far as their perceptual experience is concerned. But the focus of this experience is different from the 'embodiment' noted in the experience through machines, it is a focus which in various ways revolves around what I shall call a *hermeneutic* function. (Hermeneutics means, roughly, to interpret and usually to interpret a text.) The relation is one analogous to 'writing' (programming into a computer language) and 'reading' (interpreting the output).

Emmanuel Mounier some four decades ago already suggested that modern machinery was more and more tending away from extending our body and toward extending our *language*. And at the least the hermeneutic relation of the technical user is one which revolves, at the least analogically, around types of 'languages'. This is, of course, already a selective analogue with respect to human experience, the computer orders a dimension of possible experience. But more than this, it selects out of both the entirety of human experience analogues, not only language, but a certain selectivity within language. Again, engaging the expansive device of a continuum, our natural language abilities include the 'song' of expressive speech, filled with emotive and gestural nuances, complicated by innuendo, irony, counterfactuals, satire, often containing direct multiple and even contrary meanings expressed in between the statement and the tone in which it is said, *ad infinitum*. Out of this range of language experience the computer in its ordinary usage selects and amplifies our calculational, deductive, factoral and functional analytic experience and in its dramatic amplification which includes both complexity and speed, the computer fascinates us with its 'extension' of these experience analogues. Yet, despite the fantisied "Hal" of "2001" and the kin which he has spawned, the still equally dominant reduction of speech and language to the calculational, binary and linear remains the case.

Note that I am not trying to say that the computer either does simulate or should simulate human expressivity, I am merely noting a set of impacts which ultimately reflect back upon how we experience the computer with whatever implications that may hold.

Return to the peculiar amplification-reduction of the computer's 'language'. Not only is this 'language' a selection of possible language possibilities, but it has a characteristic shape which is its *inclination* and which provides the condition of the possibility for further structuring experience. The organization of computer 'language' is that which enhances the collection of *data*, factual information of any kind; it also enhances the breakdown of that data into *bits* or atomistic items, organized according to a system of *categories*; and in this organization there is an unfolding of what I shall call linear 'logics'. Admittedly, computer advances within this ordering have dramatic, such that even pictures which are perceived by us as gestalts can be 'built up' through an incredibly fast 'linear' process. But this highly advanced technology is not what I shall use as my primary example, rather, I want to return from the above set of factors which have largely been restricted to those who experience the computer technically, to the non-technical experience of the computer which falls mostly into the third set of categories of the experience of technology, background relations.

The non-technical experience of computers is a living with the computer as active background. In keeping with my moves from the general to the specific, I now note that the ordinary day-to-day, non-technical experience respecting computers is one which places the actual hardware away from view altogether. Computers are background in the same sense as the heating equipment is with respect to providing heat — we neither see them nor note their explicit workings. But we do, as with the heating system, experience their *results*. This is the case whenever bills arrive, when they inform us of a mistake on our income tax, or, in the detailed example I shall draw from, when they provide us with almost daily demands and results in a university atmosphere. The green and white print-outs for me as a chairman are exceeded in bulk only by memoranda from administrative offices. But while the computer machinery itself is hidden background, its results do not remain mere field as is the case of the heat produced by the heating machinery. The results often appear hermeneutically as that which is to be read and as that to which one must respond. The computer in this case has rather directly been connected to social situations — yet the computer itself is totally blind to social contexts. Its 'language', many have noted, is context blind. It does not respond to the changing, living situation, but operates in terms of a pre-set program which itself is closed to the perceptions which govern a living linguistic context. Likewise, the university computer is blind to the constantly changing social situation with results that are both negative and positive. However, I wish to illustrate this generalization by looking at a specific set of actual experiences from several sets of perspectives to attempt to isolate certain existential implications of the non-technical experience of computer technology.

In the example to follow there are two strands of issues which I wish to note. The first strand is to illustrate the inclination towards certain kinds of experience stimulated by the computer, but the second strand points to another issue at the social level, the implications for the re-organization of social power which results from the computer.

The experience itself is one which occurs within the university and relates to an aspect of record keeping, admittedly one of the social strengths of computer use with its capacity to store vast amounts of discretely organized bits of information. In this case the information is a record of a certain type of course (a category) and the grade results (bits of information.) The course is what we call a topics course, or independent study. In its actual social context at the graduate level each Department has a similar course and it is used for multiple purposes. Often it includes (a) actual directed studies on a tutorial basis, (b) research uses such as background preparation for dissertations, (c) recording of supervised teaching projects, and the like. Prior to the move to a data-based system, the Department merely kept a few sheets of records with the various grades and reported these on a semester basis to the registrar. It should be noted that there were a minimum number of sheets, but these sheets would include both all the uses of topics courses and the three different grading systems which could be used (A, B, C, grades for one use, R for continuing research use, and S, U for supervised non-graded teaching projects).

Upon the first move to a data-based system, the registrar requested that we now develop a large set of sections, one for each faculty member and a fairly large number of 'spare' sections for future flexibility. We began with 44 sections. At first, despite the fact that now the secretary must cope with a large bulk of paper — in fact greater than the mere ten courses per term of a regular order by a factor of 4 plus — the only result seemed to be a rather large wastage of paper, much of which was devoted to empty print-outs, since only a small percentage of sections would actually be used in a given semester.

But, from the technical use of the computer, this made 'rational' sense since it maximized the type of 'thinking' which the computer does, i.e. to reduce into discrete categories each type of information, and to reduce it to a linear set of ordered instances. All this was needed to straighten out the previous mosaic gestalt. But the end of the example is not yet in sight. The next thing the registrar noted was that the current program was such that only A, B, C grades were being accepted by the computer and all R or S, U systems were being rejected and had to be 'hand restored' if they were to be recorded. So, a second suggestion was made: why don't we double or triple the set of sections to 88 or 132 sections, one set of 44 for each of the three grading systems? With visions of the need to hire a student worker just to carry the bulk of paper, with fears about the quick need to buy more file cabinets for storage of records, and with the intimation that soon we would need a micro-film process to reduce bulk, we balked (along with a number of other departments). To date there is a stand-off with the registrar still doing corrections for the sake of the program.

Now, I am perfectly willing to grant that there are likely to be a number of possible solutions to this problem, most likely lying within the abilities of ingenious programmers to solve. But as I shall show in a few minutes, there is more than this behind the scene — at the moment it is probably sufficient to note that at present apparently there are more important demands being made upon programmers than straightening out the registrar's problems. But in this simple instance, we are noting both a certain inclination of the amplification—reduction result of computer technology in an extreme situation. That is, in order to order the bulk of information efficiently, speedily, and in terms of the capacities of the computer, at the other end is a price to be paid by bulk of print-out, the multiplication of discrete categories, and the sheer wastage resultant in terms of both the use of pages and of certain uses of human time.

The existential import for the non-technical user is, in this case, a negative one in two respects: first, it complicates what was a simple task; second, it reduces to a repetitive, if simple level, what was an area of judgmental competence. Furthermore, there is a social-power issue at stake here, the thesis of which is simple to enunciate: the distribution of power is now being ordered roughly according to how close one is to the source of the technical use of the computer. The Department is on the receiving end (least power), the registrar is closer to the source (more power), but the actual developer and programmer and determiner of order has the decisive power. Given both the new higher and more trivial work-load, it is no wonder that the existential import here is rated negatively.

Let us now, for the moment, step out of this example from the new uses of computer technology, and look at an earlier technological development at the beginning of the factory system to gain a distance upon a social implication of the above seemingly trivial example: During the early days of the factory system there was a deliberate strategy used by the owners and managers of industry which has as its result both the construction of new machinery and the more efficient 'control' of the work force. Bright analytic minds of the owner-managers learned to break down what was once a gestalt-skill in production, into a series of steps, each of which by itself required minimal skill. This, in turn, could be placed in a series (a linear sequence, the 'assembly line') and the now differently formed whole would result at the other end. Clearly the result was, in some cases, more speed and efficiency — but in others not. But whether or not efficiency resulted, a radical re-distribution of responsibility and control did result. The owner—manager could now (and did) *reduce* the pay of the worker, since he was doing a simpler task, and if the previously skilled type did not like it, he could be replaced by anyone who could be taught the new minimal skill in a matter of hours. The result was not only the standardization of the product, but even more the standardization of the worker. Every worker was inter-changeable and discardable. Here there was a shift of expertise, from the previous centering of expertise in the craft upon which the owner—manager had been previously dependent, to the expertise of factory organization (management skills) upon which the new, unskilled labor force was dependent. In short, the factory system was simultaneously a movement in changed technology and a movement of social change. I am, of course, suggesting that this same parallel is now occurring in some sectors and that the trivialization of work which increasingly occurs within a department with respect to records may well be part of a contemporary re-organization of skills for the sake of a new managerial elite. (I do not judge here whether the ultimate outcome will be more or less 'efficient' since that is not the point, but control, responsibility, skill will have shifted. The idle chatter that universities are more and more run for the sake of librarians [computerized], registrars [computerized] and administrations [computerized] may be more revealing than not.)

I do not intend here to imply that either the necessary result of a computerized society is this shift of power with the resultant alienation which occurred in the factory system — although I do mean to imply that that is a genuine possibility in certain cases. My point, instead, is quite different and leads to what I regard as a much deeper existential implication for the computer organized society.

(5) The existential import of computer technology is a 'world reflection'. What arises from what I have shown to this point leads in two different, but related directions. First, I have shown that technology in general, in a technologically saturated society, plays a role in organizing human experience in each of its existential dimensions. In the extension and transformation of my self-experience, my bodily self is opened to new possibilities. And while the computer plays little role with respect to these self-experiences as *bodily* experiences (at least yet) it does begin to amplify a more restricted area of self-experience, the extension of language, albeit a highly circumscribed type of language. In this role, computer technology, particularly in relation to the technical user, enters the second of the existential relations, relations with the 'other'. The technical user enters into a 'dialogue' with the computer within the technical hermeneutics of programming and reading results, and with a sense of both control (I set the program) and limit (within the limits of the computer's restricted language capacities). But both of these focal experiences of computers are exceeded by the growing omnipresence of computers and their results as a part of the immediate background of social life, the existential possibility most often experienced by the non-technical user of computer results.

The immediate result of this observation, when linked to those of the preceding parts of this paper, underlines the essentially different way in which the non-technical experience of computer technology is different in its possibilities than the technical use. To experience computer technology solely through its results within a background, even if it involves the semiactive reading of print-outs, remains at a distance as a part of the texture of the world. Technologies, in fact, today rather thoroughly texture our world as a new analogue to 'Nature' as that which is experienced. But as an analogue with 'Nature' technology may or may not be hostile.

However, when all three existential relations — machine as self-extension, machine as quasi-other, and machine as immediate world — are seen together, it may be noted that not only do they play different roles within experience, but they show a certain *telos* or direction with respect to human desire. What I am after may be expressed generally as a human tendency to 'bring close' what is far when the question of what is positive is raised. Contrarily, to leave at a distance either implies forgetfulness or even possible threat. Permit one illustration which shows what I have in mind: Recently Stony Brook received some publicity concerning what purport to be pictures of actual atoms, the first such pictures to be obtained. Now I do not pretend to know the technology required, although it contains, according to the press release, a computer enhancement technique and the use of ultra-magnification through electron microscopes. Assume for a moment that the claim is true and that this is a break-through in imaging an atom. What this picturing does, then, is to bring into range that which was previously out of range and only indirectly 'known'. We desire, in all our knowledge, to bring into view that which is far, ultimately we want it 'in the flesh'. But in terms of the series of existential relationships previously described, this is to bring what was in fact 'hidden' within the far background of the world into what at the very least lies between a hermeneutic and an embodiment relation. The movement was one *from* background towards being experienced *through* the machine. The ultimately preferred location lies closest, in the embodiment relation.

This implication of a preference or telos towards closeness over distance permeates what I have noted is the essentially different relation the technical as compared to the non-technical experience of computer technology entails. But there is a more important point to be made. In human existential relations with a world, one may note historically that there are interesting and often dramatic changes in the way humans interpret both their world and themselves. This movement is one which I earlier called, without definition, a 'world reflexivity'. It may be described as follows: humans interpret their world in terms of some focused interpretation, in terms of what one may loosely call an 'image'. But because humans are also existentially and necessarily related to what they perceive as their world, they 'bring it close' so that ultimately they also interpret themselves in terms of their world. Put most simply, the 'image' of the world expressed as the interpretation of what world is, is reflected back, ultimately, into a self-interpretation in the process of 'bringing close'. I interpret myself in terms of my world.

It is here that I wish to conclude by looking at a depth existential implication concerning the human experience of computer technology. And this implication perhaps effects in some ways those who are closer to the technical experience of computers than those farther away, although the same movement may be seen at both ends of the continuum. First, take what is a common interpretation in a strongly negative sense of what I am pointing to in terms of the non-technical experience of computer technology.

Students subjected to long lines at registration, who must not only fill out sheets to be scanned, but who are treated like child—idiots by those who check over the sheets for errors, and who then, in spite of checking endure the guilty-unless-proven innocent results of sometimes ineffective procedures, begin to resent what they perceive to be a 'cause' of their predicament. The signs seen in the student movements to the effect that "I am *not* a number or computer card, do not fold, mutilate or spindle" are expressing negatively the tendency for us to interpret ourselves according to a dominant 'image' or relation to what is perceived as world.

At the other end of the continuum, there is what is taken by those who express it as a positive interpretation, a functionally similar belief concerning artificial intelligence. Does the computer genuinely simulate or model our thinking processes? Is the child really 'pre-programmed' to learn a **language?** etc. Both of these moves reflect the same tendency, the tendency to form an 'image' of the world according to frequent or dominant experiences, in this case the technologically saturated experience of contemporary life, and to ultimately reflect that back into one's own self-interpretation. This is a possible depth existential implication of computer technology for humanity.

I shall not here address the other side of the issue, but I shall note by way of an observation and a question that the technological transformation of experience with its ultimate result — a 'world reflexivity' — needs to be very carefully examined if my observation that one form of that transformation is the amplification— reduction selection. I noted earlier that the amplification is always what stands out dramatically and obviously, but that the reduction may be 'forgotten'. Yet both belong equally to the transformation which occurs. It well may be that ours is the age of 'calculative thought' as Heidegger termed it. But if this is so, it is equally but one possible way of being and is, I would dare to say, no less 'anthropomorphic' than any other way or era. And what is needed as much as the attention to the possibilities of amplification is the needed critique which reminds us of reduction. This is because the existential implications of computer technology, like all existential implications, are essentially and necessarily ambiguous.

CHAPTER 6

TECHNOLOGY AND THE TRANSFORMATION OF EXPERIENCE

Underlying much of what we usually take for granted within experience there are various subtle and hidden structural features which a phenomenology seeks to make explicit. But the situation becomes even more complex when in the course of experience some artifact is used. In the analysis to follow I shall try to demonstrate certain features which technology in use implies for human experience. My thesis is that any use of techology is *non-neutral*. However, non-neutrality is not a prejudicial term because it implies neither that there are inherently 'good' or 'bad' tendencies so much as it implies that there are types of *transformation* of human experience in the use of technology.

What is interesting for the phenomenologist is the search for and hopeful discovery of structural or invariant features to be found in the process. In what follows, I hope to demonstrate through a descriptive analysis that there are such invariant structures in the use of technology. The limits I have set here will be recognized by some to bear certain resemblances to Martin Heidegger's "tool analysis" in *Being and Time*. But I also will develop those similarities somewhat more intricately than his historic analysis.

I. EXPERIENCE TRANSFORMED

Epistemology is clearly one of philosophy's central concerns. In the choice of examples which can serve as paradigms for other technologies, I have selected instances of technologies which bear directly upon the knowledge situation.

I have chosen to examine a quite narrow problem set which I shall call 'knowledge gathering' technologies, or simply, instrumentation. Furthermore, I have limited these to examples from what I shall call 'visual technics', or instruments which yield some type of visual result.

Methodologically, I shall employ a *phenomenological descriptive analysis* which relies upon an adaptation of Husserl and his concept of intentionality. I shall limit myself, accordingly, to an analysis of instruments *vis-à-vis* intentionality, or put simply, to the role of instruments in their experienced use.

The analysis will have two variables to correlate. The first variable is perception — in all the cases to be examined a perceptual act is involved as a necessary condition for the gaining of the knowledge which is gathered. Thus I shall be concerned only with cases which involve perceptual situations explicitly.

The second variable feature of the analysis will involve the experienced use of some material artifact in special cases of knowledge gathering, in short, the use of an instrument which I take to be an instance of a 'technology'. It should be obvious that the second variable covers a special set of cases, presumably narrower than the field of all perceptual situations.

With both variables, I shall restrict myself to visual examples, although I shall regard these as symptomatic of wider issues in the employment of instrumentation in knowledge gathering. I shall call the use of instrumentation in a visual situation, *visual technics*.

II. THE PERCEPTUAL SITUATION

Although the analysis to follow is not necessarily intended to apply only to scientific activities, I have chosen examples from what I shall call *observational* activities related to science because I think these most dramatically illustrate the issues which arise in the human use of instrumentation. Observation, then, will be seen to entail (a) a perceptual activity, here visual, and in special cases (b) a perceptual activity which also includes the use of a technological artifact, an instrument.

I begin with what I shall call a *direct perceptual situation*, that is, one which does not include the use of any artifact.[1] Such a situation, analysed in terms the structure of intentionality (as employed by Husserl), may be schematized as follows:

$$\text{Human} \longrightarrow \text{World}$$

$$(\text{Observer} \longrightarrow \text{Environment})$$

The minimal phenomenological interpretation of this state of affairs would include the following elements: (1) The human experiencer, general case here illustrated by the visual observer as the special case, (2) is intentionally engaged, here interpreted as a directed and referential activity, (3) with an experienced World or aspect of the World, here restricted to examples of what is visually experienced within the environment. In Husserlian terms, the terminus of the intentional aim is called the *noematic* correlate; the activity of experience, the experiencing, is called the *noetic* correlate. The overall situation is the structure of intentionality, an instance of 'all consciousness is consciousness of _____'.

I assume here that intentionality has the general features described in the phenomenological tradition: intentionality is directed, is structured such that there is a referential aim which may or may not be fulfilled, exhibits itself as having both a core or focus and a fringe or horizon, etc. In visual terms these features may be quite literal. A visual intentionality is one in which there is an identifiable focal structure located within and centered in a visual field. Its *shape*, then, is one which may be described in terms of a focal-field ratio in which those objects which fulfil an intentional aim occur as figures located within a field or background.

For purposes here, direct perceptual situations are those in which observation occurs through a visual engagement directed towards the visible environment. Knowledge is gathered when the fulfillment of, or terminus of, the intentional aim is attained.

In more general or ordinary terms, I take it that direct perceptual situations are quite ordinary situations for knowledge gathering and have so functioned since the origins of humanity. However, in some subsequent time, humans invented and began to employ instruments in gathering knowledge, not the least of which were the investion and development of optical instruments which were closely associated with the rise of the modern scientific era.

I shall argue that the use of such instruments modifies and transforms – in quite definite ways – the knowledge gathering situation. Speculatively and historically, I believe that modern science is nessarily *embodied* in its instruments which function as a necessary conditon for its knowledge gathering. I shall seek to exhibit that result through the examination of how instruments have modified the intentional situation.

III. MEDIATED PERCEPTUAL SITUATIONS

In terms of the experienced use of instruments, it will be noted that a new factor occurs within the perceptual situation, the technological artifact. I shall call states of affairs in which an instrument is used, *mediated perceptual situations*. But in this context, the question is quite specific – what modifications occur with respect to experienced uses interpreted phenomenologically? The aim will be to isolate any *invariant features* entailed in mediated perceptual situations as compared to direct perceptual situations.

Pre-phenomenologically it is possible to note that in the employment of visual technics, one simply 'looks through' some optical instrument 'at' some target and either succeeds or not in seeing something. In this case the instrument 'mediates' what is seen in some way. It is possible to schematize this state of affairs in terms of the previous model of intentionality:

Human —— artifact ——World

(Observer —— instrument —— Environment)

Here the specific class of cases of instrumental mediation are those in which the perceptual act is mediated by means of an optical instrument in order to get at some feature of the visible environment. But in terms of an analysis of intentionality, this is, in effect, to have taken the instrument *into* the intentional correlation. It occupies what I shall call *mediational position* and much of the task of the analysis is to isolate the features which emerge within intentionality with the inclusion of an artifact in mediational position.

I shall argue that the instrument changes certain aspects of both the **noematic and noetic correlates** and attempt to exhibit a *structure* or invariance entailed in the use of instruments.[2]

One strategy I shall employ in the analysis, itself descriptive in aim, is to contrast the features of both noematic and noetic correlates as exhibited in direct as compared to mediated perceptual situations. Also, prior to the descriptive analysis itself, it may be necessary to enter one qualifying caveat: much debate which can lead into dead ends here can be forestalled if it is seen that 'theory laden' perceptions can be dealt with by taking some level of 'theory lading' as a constant. In short, much useless debate about perception and knowledge gathering ends up in confusions about levels of background information. It is quite obvious that there is a vast difference between, say, a novice observer who lacks background, and the expert in the field. Thus the novice bird-watcher may report merely that he saw a 'brown bird sitting in the field', whereas the expert will report that the bird is a 'Dupont's lark, distinguished by its slightly down-curved beak, brown lines along the breast, etc'. Now, the bird *has* those features mentioned – they are not invented by the expert – but they are noticeable as important to the 'theory laden' perceiver. However, this background information can serve as a constant, in which case we shall not confuse levels of observational expertise. I shall assume here only highly trained observers.

It may now be expected that possibly interesting epistemological problems may emerge by the examination of instrumental technologies.

IV. A PHENOMENOLOGY OF INSTRUMENTATION

The aim of the following analysis is to isolate and take account of (a) the role of instrumental transformations of the perceptual situation, (b) the exhibition of invariant structures of instrumental mediation, and (c) to locate, thereby, epistemological implications entailed in the use of instruments.

I begin by noting a few features, both noematic and noetic, of direct perceptual situations. It may be said that noematically, what usually occupies a visual act, is a field filled with what I shall call 'middle-sized' objects. As a display, these are arranged above, below, in front of, behind, alongside each other in a three-dimensional depth field. Although the features which may be displayed vary enormously, it should be noted that there is a range to the field of what is visually possible regarding size, distance, the ratio of clarity to indistinctness, etc. Although I shall not now take the time to fill in the structural properties of the visually noematic field, I shall take it that it is possible to recognize most of its features if mentioned as familiar to ordinary experience. However, I also wish to take note of one *reflexive* feature of such direct perceptual situations.

Whatever is displayed is *reflexively* related to actual bodily position. Whatever I see appears as far or near, as being distinct or indistinct, etc. with respect to my relative bearing within the environment as such. I may schematize this reflexive situation of bodily position, intentionally as:

$$\text{Human} \rightleftarrows \text{World}$$

To phenomenologically interpret this situation, I note that the noematic correlate, World, refers back, albeit implicitly, to the position of my body within it. Conversely, my actual bodily position serves as a noetic condition of the possibility of what I see as I see it. Thus, if an object is in the far distance, it may appear as indistinct (unless of a large size) and regardless of size its micro-features will remain either indistinct or not noticeable at all. For such features to change noematically there must be a noetic change in my relative bodily position. I take it that such ratios and variables are indices to *structures* of visual intentionality. What is initially important to note here, however, is that it is (a) in the correlation of noema and noesis that such structural features appear, and (b) that bodily position is taken account of implicitly, rather than directly. It is reflexively detected.

I now begin the examination of technological transformations of this invariant set of direct perceptual structures. I shall here employ a set of variations upon visual instruments in what would be recognized as a typical phenomenological exercise in the use of free variation, the aim of which is to isolate essential features or structures which are to be exhibited through the variations.

TRANSFORMATION OF EXPERIENCE

My first example of a technologically mediated perceptual situation comes from the use of a telescope. Telescope use is an instance of the previously characterized mediated intentional arc. In this case let us imagine that the user is directing his gaze at the moon and that he is the first human user of a telescope. The formal situation is now one in which the telescope is placed in *mediational position:*

Observer —— telescope —— Moon.

The instrument now occurs within the realm of intentionality and one may describe the noematic and noetic features, both alone and in comparison with a direct perceptual situation which here would be a matter of looking at the moon with the naked eye.

Now, in both the direct and the mediated perceptual situations, the terminus of the intention is the moon and, in both situations, it may be called the *object* of experience. But its appearance changes — and if the telescope is good and of a high power, it may be said to change *dramatically*. Suddenly, there appear to be mountains and craters and valleys on the moon, and in the imagined situation here since the viewer is the first person to look through a telescope, it is apparent that some kind of *new knowledge* has been gained. In some sense yet to be more fully determined, the telescope has functioned as a *condition* for the gaining of this new knowledge.

The change in the appearance of the moon, phenomenologically a noematic change, contains far more than I have noted initially, but tactically because the theme of the analysis is upon instrumental use, I shall begin immediately to turn to the role of the instrument. But when I do this, something quite curious occurs. Although it may be said that the instrument *appears*, in some sense, in the experience, during use it is clearly *not* the object or terminus of the visual intention. It is, rather, the *means* of the experience. But in being the means of experience the instrument itself may be said to 'withdraw' from any thematic obviousness.[3] Indeed, this part of its function as a *condition* of attaining the new knowledge.

A finer descriptive analysis of this phenomenon can show more fully what occurs. In use the 'appearance' of the instrument is such that it is, at most, a background feature of the overall visual situation. If the telescope is serviceable and functioning, it becomes semi-transparent in the use situation. It is only by means of this transparency that the moon is seen at all. Thus, experientially and in terms of the intentional arc, it becomes possible to characterize the instrument as taken into the observer's now extended self-experience. The instrument is experienced in use as quasi-extension of the acting observer. I shall characterize this state of affairs as:

(Observer—instrument) ⟶ World

Here the parentheses indicate (a) a state of the instrument-experienced-as-semi-transparent and (b) functioning as an experienced extension of the observer, i.e., as a partially symbiotic unity. This use situation does not exhaust instrumental possibilities, but it is only with situations of this sort that I shall deal here.

The partial symbiosis of observer—instrument, signified by the dash within the brackets, is highly enigmatic. On the one hand it may be seen immediately that the transparency of the instrument is a functional condition for the attainment of new knowledge, yet on the other hand its function is difficult to probe for significance precisely because its positive qualities are those of 'withdrawal' and 'transparency'.

I have characterized this situation as one of a *partial* instrumental transparency and as one of a *partial* symbiosis. That is the case in spite of high efficiency and good resolution. The instrument never completely 'withdraws', but leaves a vestigial presence. And although I shall return to the noematic clues which are indicative of this aspect of the phenomenon, the mere noting of the superiority of an actual view of the moon from, say, a circling satellite, is obvious enough.

The instrument does have a fringe or background presence. Yet this may be seen as an 'imperfection', in that once it is seen that 'withdrawal' or transparency is part of the condition of new knowledge, it is possible to fictionally project a 'pure' or perfect transparency as an ideal limit. But here there is occasion for two very serious confusions: I have noted that the instrument is in some sense a condition for the gaining of new knowledge. I shall now argue that the projection of a fictionally 'pure' transparency has as its internal logic (a) a rejection of instrumentally mediated knowledge as necessary, and (b) it functions so as to be equivalent to an ideally *direct* perceptual situation. There is in this confusion a 'leap' between existential and conceptual possibilities which is, phenomenologically, a gross category confusion.

There are two reasons which serve to undercut the confusion. First, transparency alone is not all that is involved in an instrument's being a condition of the possibility of new knowledge. Some transformational property is also required. In short, some *difference* between a direct perceptual situation and a mediated perceptual situation is required in order to have a condition of the possibility of new knowledge. Secondly, although more subtle to note, there may be detected some actual phenomenological difference in any instrumental mediation such that it must be essentially characterized by such differences.

A temporary change to a variant upon telescopic optics may indicate these reasons. Suppose, now, instead of a telescope, that our observer views the moon through a very clean and well made plate glass window. We might imagine for this case that this plate glass has the highest actual degree of transparency within current visual technics. But in such a case, while the moon continues to be seen, it is seen almost as seen by the naked eye — its mountains and craters remain as undetected as previously and the plate glass remains so isomorphic with the direct perceptual situation that no new knowledge is gained. An equivalence to a direct perceptual situation through 'pure' transparency gains nothing new.

This is not to deny that transparency is one of the elements of the visual condition for knowledge. Vary the telescope example by imagining either opacity or poor resolution because of grease-smeared lenses and transparency will be seen to be essential. But it is not a sufficient condition. For that, some difference between a direct perceptual and a mediated perceptual situation must obtain. In this case I shall call it a transparent-*magnificational* property. Here it becomes more obvious that the instrument, precisely because it is transformative of direct perceptual situations, may be a sufficient condition for a new knowledge.

However, before returning to explore the transformational features of instrumental mediation, I wish to take account of a few very subtle features of even the 'pure' transparency of the plate glass. And it is here that the constant of a highly trained observer must be noted: phenomenologically, there are notable, if subtle, differences between the noema seen with the naked eye and those mediated through the plate glass. With plate glass it is often the case that there is residual 'back-glare' which ever so slightly interferes with its transparency. There is also a certain very slight 'flattening' effect upon the object seen; that is: some slight loss of depth. Such changes in the noema indicate, even if not always detected by the less critical observer, a change in noesis. These same features become more dramtic in optical instruments.

I have now indicated that there are at least two co-present factors for the instrument to be a condition of the possibility of new knowledge: *transparency*, however partial or close to an ideal, and some kind of *transformation*, here called magnification. And just as there is a confusion between the conceptual and the existential with regard to transparency, there is an equivalent confusion with regard to instrumental transformations. The second confusion is one which relates to the interpretation of instruments as purely *neutral*. Its implicit logic assumes that even if instruments were necessary for the attainment of new knowledge, their use is merely neutral and in no way effects what is known. The mistake here is isomorphic with the mistake made concerning transparency. An ideal transformation or magnification via an instrument, would deliver the thing itself just as it would be for a direct perceptual situation.

That this is not the case with actual instruments is all too painfully apparent to their users, but what is more important to demonstrate in this situation is what is involved in an instrumental transformation. I shall attempt to exhibit, phenomenologically, that the transformational capacity of instruments in use entails a two-sided phenomenon, that simultaneous and co-extensive with every magnificational capacity there is a reductive effect. I shall try to show that this *magnification—reduction* is a structural or essential feature of instrumental mediation and is the basis for the non-neutrality of instrumental use.

V. AN ESSENTIAL MAGNIFICATION—REDUCTION STRUCTURE

To exhibit the essential magnification—reduction structure as an invariant of instrumentally mediated perceptual situations, I shall have to utilize both noematic clues to noetically implicit bodily situations and a comparison of direct and mediated perceptual situations.

That the magnificational transformations of visual technics is dramatic and stands out, goes without saying. The excitement which accompanied the initial discoveries of moons, planetary rings, and other celestial phenomena, and the equivalent surprise at the new knowledge of cell structures microbes and other creatures through the microscope easily stimulated early scientific imagination. This drama, however, often obscures the hinderside of instrumental transformations which I have here called the reductive dimension. It is not out of any negativity towards instrumentation that I shall emphasize the latter more hidden and subtle side of instrumental mediation.

I begin with a partial account of noticeable noematic differences between directed and mediated perceptual situations: (a) In a direct perceptual situation, here viewing the moon, the object is always located in some *expanse* or field. The moon is surrounded by the night sky. But through the mediation of the telescope, the expanse is changed. At high magnification, for example, the moon is no longer 'in the sky' at all, but is before one through the telescope, its field perhaps defined by the vestigial presence of the instrument, the back circle of the telescope tube, as if the moon is seen through a 'hole'. More can be made of this, but I shall remain satisfied if it is merely noted that an instrumental mediation changes the *expanse-variable*. (b) The same type of change occurs with what may be called the *depth-variable*. Here the moon example does not serve as clearly as another variant, the microscope, to which I shall turn briefly. But even with the moon, it may be noted that its depth appearance changes with the use of a telescope, and the change is more than a matter of apparent distance. It may be seen, for example, that the mountains and craters have depth (particularly in cases where shadows accentuate this) but the depth is noticeably a lens-mediated focal depth which is fixed and 'shallow'. The microscope makes this more apparent — if one views 'wee beasties' in fluid, the focal depth is often very shallow so that what is below the correctly focused 'beasty' may not be seen at all or if seen, seen only as very fuzzy. Noematically, both expanse and depth variables change when mediated through optical instruments, and this transformation is more than a matter of presumed visual distance. (c) I may combine these observations as an index of one instrumental invariant: the noematic appearance mediated by an optical instrument is one marked by a fixed *near-distance*. By isolating this feature of a fixed near-distance, I may, moreover, note that this phenomenal distance is the *same* for both telescope and microscope — the object is there before me in the fixed near-distance of the focal plane — in spite of the very markedly different distance of my bodily position *vis-à-vis* moon and microbe. But here I anticipate one of the noetic and implicit variables as well.

Already here, it may begin to be obvious that the magnificational gain is accompanied by a subtle but noticeable reductive transformation of the object. This underside of the instrumental transformation, however, becomes clearer if we follow the correlated noetic clues for a moment: (a) I have indicated that all visual acts include some implicit bodily involvement which may be reflexively taken account of. The same applies in this example. What the instrument does, in use, is to change the *apparent* bodily position *vis-à-vis* the object. In the use of the telescope this *apparent* position is 'closer' to it (as if in a space ship only a few thousand feet above the surface). (b) But this apparent noetic-bodily position is not total. It conveys with it a certain sense of **irreality** or of 'abstractness'. It is 'as if' I were closer to the moon, but I do not fully feel myself 'there'. Thus the **bodily** involvement in a mediated perceptual situation displays a feature which is different from the more immediate and global involvement noticeable in a direct perceptual situation. (c) Here the noetic index may be reversed to detect a subtle noematic effect as well. The near-distance, previously noted, of an instrumentally mediated object may also be characterized as partially 'abstract' or lacking a certain global concreteness. I contend that this feature marks *all* instrumentally mediated noema, and that it is a clue to the reductive side of the magnificational capacity of the instrument.

Phenomenologically, of course, it is possible to account for this effect by noting that direct perceptual situations are ones which involve global bodily involvements. Of course in the use of an instrument, there continues to be a bodily involvement, but it is always slightly *displaced*. At the least, the use of the instrument itself displaces a part of our bodily involvement with the object. And here we find a clue to one further feature of instrumentally mediated objects. The object mediated by an instrument may appear as a sensorily *reduced* object. One instrumental capacity here, is to 'analytically' reduce a perceptually rich object to a *mono-feature* of itself. By this I mean something quite simple, but also something which contains a much greater set of implications than is usually taken to be the case. An optical instrument gives us *only* a visible object or gives us only a visual slice of the object. In this sense the optical instrument is mono-sensed (implicitly, I am also suggesting that humans are never mono-sensed as a bodily noetic correlate).

Now this aspect of instrumental mediation does not exhibit itself as well with the optical variants used so far, the telescope and the microscope, precisely because the normal 'distances' of humans from the moon and the microbe have been so great. Until very recent times, we have never experienced a full face to face moon which is touchable, tasteable, smellable, etc. But hypothetically, we have known for a long time that the moon does have these rich perceptual characteristics. And because we implicitly take global bodily engagement as a taken-for-granted norm, even through instruments, say a telescope with very high magnification, we may have noted that what appeared as fine powder on the moon *must* feel like dust. For globally experiencing humans, even a reduced visual result carries a fringe of global experience. But this is not a fulfilled intentional aim through a merely optical instrument. Thus its reductive limitation to a mono-dimension may be quite clearly noted as a phenomenal feature of its use.

I do not claim here to have phenomenologically exhausted what is exhibitable in mediated perceptual situation, but the fixed near-distance of a mono-dimensioned and partially 'abstract' or irreal perceptual object as the noematic correlate should be sufficient to have demonstrated the reductive dimension of the instrument's magnificational structure. Thus I now return to the positive aspects of instrumental mediation.

The positive dimension, is of course the magnificational dimension. If I describe this phenomenologically, I note that the instrument in its capacity to bring to near-distant presence is selective. It brings what I shall call mirco-features into perceptual range. Now in the case of the telescope these have been such features as would appear to be micro-structures only in relation to the great distance between the earth and the moon. The moon at its great distance, has a micro-structure, its mountains and craters. I have noted that this instrumental capacity entails a change in *apparent* bodily position noetically, but it is now actually possible to be in such a position *vis-à-vis* the moon. Nevertheless, historically, much of our knowledge of the moon was gathered telescopically — the telescope was a condition of the possibility of this knowledge within that context. However, with the microscope, we reach a different level of micro-structure altogether. Here magnification is so great that the only way we could come in contact directly with the microbe would be to have the entire size of our body changed. Thus here a micro-feature is something which becomes present *only* by means of the instrument. *For embodied beings, embodied perceptual extensions are a necessary condition for the expansion of perceptually gathered knowledge of micro-features.*

This is, again, historically what characterizes modern as contrasted with much ancient science. Modern science is *technologically embodied*. But while I have shown that the instrument serves as a condition of the possibility of certain kinds of gathered knowledge; and I have begun to exhibit the non-neutral transformational structure which characterizes the way in which this knowledge occurs by means of the instrument I have yet to show what this transformation implies for a proper understanding of a changed epistemological situation.

VI. INSTRUMENTAL 'INTENTIONALITIES'

It is the *difference* between direct perceptual situations and mediated perceptual situations which yields the new range of knowledge possible by way of instruments. This difference is one of utilizing the non-neutral transformational structure of instruments. But this is to have phenomenologically isolated what I shall now call instrumental 'intentionalities'. I place intentionality in this somewhat metaphorical sense in quotes because I do not want to mystify about machines, nor do I impart to them anything like consciousness as such. But insofar as phenomenology claims to isolate and describe what may be called the *shapes* of intentionality (experience), the same may be seen to apply to the extensions of intentionality through instruments.

The difference is that all instruments have *differently* shaped 'intentionalities' which expose precisely those aspects of the world which have hitherto either been overlooked, taken as unimportant, not known at all, or even totally unsuspected. And it is precisely in the *difference* that much of what we consider new knowledge falls out. I shall use one final example to exhibit this feature: not all visually mediated results today are as close or as isomorphic with human vision as the simple telescope and microscope. The example I have in mind here is the use of infrared to reveal what was previously unseeable altogether in directly perceived situations.

Suppose now that we have a special instrument which projects infrared upon a certain scene. Suppose that this projection also includes a device which makes the result of the infrared projection visible to us (as in the case of infrared photography).[4] In this case not only is the vast difference between human and instrumental 'intentionalities' apparent, but what is perceived in this case is not only different in degree of micro-structure, but in kind. The infrared projection may make perceivable, for example, regions of disease in plants which were undetectable in any direct perceptual situation. Here it is by means of the different instrumental 'intentionality' that new knowledge is gathered. But it continues to be gathered by being brought into the near-distance of the transformed perceptual field of the instrument.

Thus, not only is the instrument a necessary condition of the possibility for some kinds of knowledge, but this knowledge is gained by means of the non-neutral difference which occurs in an instrumental 'intentionality'.

VII. A FABLE: AN INSTRUMENTALLY CONSTITUTED 'WORLD'

I may be able to consolidate the previous analysis by turning to an imaginative example. There is a fictional account well known to many mathematicians and philosophers of the intersection of a two and a three dimensional world.[5] In the story the beings who live their lives in a two dimensional world — points, lines and other geometrical figures restricted to a single flat plane — find themselves highly puzzled by the apparently absurd intrusions into their world by beings who inhabit three-dimensional space. Thus, for instance, a sphere which might happen to suddenly appear in two-dimension land, would appear as a point, widen into a circle, again decrease to a point and finally disappear. The two-dimensional beings, naturally, have a difficult time accounting for such phenomena, whereas, conversely, the three-dimensional beings are perplexed about the limitations of the two-dimensional beings.

One might suppose that there is a simple parallel between direct perceptual and mediated perceptual situations. Imagine two 'worlds' in which in one all one knew was in terms of direct perceptual situations and in the other all one knew was in terms of instrumentally mediated perceptual situations. In this second 'world', continue to imagine that the only instruments are optical ones. It should be quite obvious that the appearances of these 'worlds' would be quite different.

In the totally instrumentally mediated 'world' whatever would appear would do so in terms of the relatively 'flat' near-distance of an optical plane. Although the size of objects might change with the use of different lenses, the object-distance would never change. Moreover, whatever objects there were in this mediated 'world', would appear within the limits of visually reduced form such that they could not be touched, smelled, tasted, but would remain mono-dimensional in appearance as totally visible beings. Such a 'world' would retain the sense of 'abstraction' and near-distances previously noted.

Imagine, then, the perplexity of the inhabitant of the direct perceptual 'world' who entered instrument land. He would find himself confronted with what to him appeared to be highly reduced, untouchable, constantly distant entities which always seemed to be partially 'abstract'.

Here the difference between a direct perceptual situation and the instrument mediated situation is parallel to the situation which obtained between two and three dimensional beings of the *Flatland* fable. For fully-sensed beings, the richness of multiple dimensions should be apparent. I am suggesting in this fiction that there is a latent and implicit 'preference' or human desire for a global awareness of what is other, in contrast to the partial restrictions imposed by the use of instrumental mediations. This desire often supplies a direction for the development of instruments themselves.

Of course it is possible to imagine a fully instrumented world in which instruments embody other sensory dimensions. And it is precisely the genius of contemporary science that it has carried the development of such instruments into dimensions other than the visual. Much of the new knowledge gained of the soil of Mars, for example, has been by means of instrumental analogues of kinesthesia, taste, touch and smell. But even if each instrumental analogue were developed and then all of them coordinated, I suggest that the transformational structure would remain.

Behind this development and sophistication of instruments lies the deeply held but usually unexpressed belief that the most genuine knowledge is ultimately one which is given to a globally full perceptual situation. Ultimately, we want to be 'face to face' with the things themselves'.

However, with respect to instrumentally mediated situations, there is another side of the fable as well. Even the inhabitor of the instrumental 'world', were he to suddenly come into the 'world' of direct perception without his instruments, would face a new and different counterpart perplexity. He might well wonder what had happened to his well known mountains and even craters on the now suddenly far distant moon and he might further wonder what happened to the microbes and other familiar life within what now appears as a merely transparent liquid. For unlike the *Flatland* fable, the instrumentally mediated world yields, however sensorily reduced, a realm of micro-features which except for instruments would never come into the horizon of the direct perceptual being. There is, in short, a correlate to the 'naive realism' of the direct perceptual world and it is what might be called 'instrumental realism'. Once instrumentally presented what was previously unknown offers itself as a 'thing itself'. Moreover, for the bulls in chinashops which we are bodily with respect to the microworld, the instrument is the necessary condition of the possibility for our peeking at that world.

There remains one final word for the inhabitor of the direct perceptual world, and that is that in whatever form an instrumental mediation may occur, it nevertheless must make a reflexive reference — direct, mediated or indirect — to direct perception. In all the cases examined one might point out that certain invariant features of direct perception remain constant even in and through an instrumental mediation. Thus the instrument may bring a micro-feature to presence, but to do so in the case of humans, it must present the phenomenon to our visual field, and to focal location within its core. Subtly, the very expanse of the human direct perceptual field is a condition of recognizing the limitations of the instrumental mediation.

Thus for the gaining of knowledge in the contemporary sense of a technologically — embodied science, one can say of instruments that one can't entirely get along with them, but neither can one get along without them.

NOTES

[1] I shall not enter here any question about whether or not the human body is an 'instrument', but shall hold that 'I am my body' as a basic assumption of the paper. All that is important for the case here is to note that *if* the body were an 'instrument', it is indeed a very different one than those we use.

[2] Rather than use the confusing Husserlian 'essence', I use structure to describe invariances.

[3] In part, this analysis is dependent upon Martin Heidegger's so-called 'tool analysis' in *Being and Time*. It was he who first descriptively analysed instrumental 'withdrawal' in use. But I hope it will be seen by experts in this domain that the analysis here goes into detail not developed by Heidegger.

⁴ There may be an instrument such as I imagine here — perhaps a type of sniper-scope uses infrared. I do not develop the obvious use of infrared photography here because this use is not of the type which occurs within the *embodiment relations* of this paper.
⁵ See Edwin A. Abbott's 1884 *Flatland: A Romance of Many Dimensions*. (Dover).

CHAPTER 7

VISION AND OBJECTIFICATION

Several years ago there appeared an essay which was the programmatic opening for a series of studies in the phenomenology of sound.[1] In setting the context for that investigation I made note of a 'visualist' tradition within the history of philosophy. From Heraclitus who declared that "eyes are more accurate witnesses than ears," through Aristotle's claim that "sight is the principal source of knowledge," into the modern era opened with an en-*light*enment, the dominant metaphors for thought have remained visual ones. Often this "visualism" has been accompanied by a lack of attention to the other sensory dimensions with a resultant reductionism which may be called a *reduction* to *vision*. Today I would maintain that the dominant strands of philosophical thought still continue the 'visualist' tradition where vision stands as the root metaphor for the clarity, rigor and distinctness desired by the philosopher.

However, with the appearance of a discussion article on the research into sound, a second dimension of philosophical 'visualism' needs to be discussed. Walter Robert Goedeke's discussion praises a certain version of 'objectification' when he claims that, "the anguish of discovering that one is *nothing but an object* in the world is a necessary step out of an innocent Eden into self-creativity (italics mine)".[2] In the process of discussing objectification, however, Goedeke implicitly accepts the long tradition which associates *vision* and *objectification*. He links, for example, the rise of objectification with visual phenomena such as 'soundless' logics and most particularly with the printed word which is a visual embodiment of language.

I shall not here address myself to the question of the value or disvalue which objectification may have for human self-awareness, nor am I prepared to undertake here what is even more intriguing from a phenomenological point of view, that is the question of the de-centering of spoken language by writing which calls for its own phenomenological investigation.[3] But the question which I do wish to address is the crossing of the first 'reductionism' of 'visualism' which reduces the paradigm of thought and experience *to* vision with a *second* 'reductionism' which then is a *reduction of vision* itself. The second reductionist move is that which reduces vision to an *objectifying gaze*.

If the first reductionist move begins to close off avenues to the understanding of the fullness of experience in all its dimensions, the second complicates the closure by hiding what we do and can see, the essential possibilities of vision. In short, I wish to 'demythologize' the tradition which links vision with objectification and show that a phenomenology of vision finds 'more' in vision than the presence of 'objects'.

It must be admitted that the sources for a relation between vision and objectification do lie as one set of possibilities for visual experience. When I look, I look 'outwards' and the objects which appear to me appear as *at-a-distance* and as *spatialized*. Thus the object upon which I gaze is 'out there', 'apart from me', 'at a distance'. Moreover, it presents itself as having a body-outline, a *shape*. Ordinary objects may show themselves as *surfaces* colored and extended. The object carries with it an Otherness. Such is a central and ordered *possibility of vision*. But this description is not adequate phenomenologically even if at its shallow level it is 'apodictic' so far as it goes.

A more adequate phenomenology of vision must seek to discern the essential possibilities of vision stretched to limits. Moreover, such a phenomenology should approach the above description with a suspicion which demands a guard against too quick closure. This suspicion lies in *epochē* which 'brackets' the possible 'reality assumptions' which may lie embedded in the above description such that a superficial apodicticity may pose for adequacy. To accomplish this purpose, phenomenology 'horizontalizes' a full continuum of visual possibilities to see what they may hold as implications for the structure of vision.

This 'bracketing' of implicit 'reality assumptions' may be illustrated by following an imaginative example which 'possibilizes' the broader visual situation: Imagine two seers, one of whom we shall call a 'cartesian' seer, the other a 'spiritualist' seer. Both are assigned the task of 'observing' a set of 'tree-appearances' under a set of varying conditions and then report on what the tree 'really' was like. To make the story short, the 'cartesian' seer returns from his observations with a very accurate description of the tree's color, shape of leaves, texture of bark and characteristic overall shape. Upon interrogation, however, we find that out of the range of conditions under which the tree appearances occurred the 'cartesian' seer has clearly chosen as *normative* only those appearances of the tree in the bright sun on a clear day. His 'clear and distinct' tree, characterized as essentially an 'extended, shaped, colored configuration' is a 'cartesian' tree which appears best under the light of day, all other conditions being dismissed as less than ideal for observations.

The second, 'spiritualist' seer returns with quite a different description. His tree is one which emerges from an overwhelming nearness of presence and is eery in that it bespeaks its druid or spirit within. It waves and beckons, moans and groans, advances and retreats. Upon interrogation it turns out that his *normative* set of conditions were those which occurred during misty nights and windy mornings of the half-light of dawn, or when the tree appeared as a vague shape emerging from the fog or as a writhing form in the wind. His tree is a 'spiritualist' tree in which the quiet sunny day fails to reveal the 'inner' tree 'reality'.

We now have to try to reconcile the seers and the arguments which they have with one another which fail to convince either one. We phenomenologically discern that the normative appearance conditions which govern what is taken as seeing are very intricately involved with two sets of implicit 'reality assumptions' about trees and their 'natures'. The 'cartesian' seer believes that reality is clear, distinct, extended, colored and shaped — and those appearances which tend to disconfirm this belief are also ranked as 'distorted, befogged, unclear' appearances, all of which have to be rejected as a deficient mode of seeing. Contrarily, the 'spiritualists' seer holds that bright sun appearances are misleading as they mask and hide the true animated 'reality' of the tree which only shows forth under the conditions of mist and wind and rain, thus bright, daylight appearances are also contrarily rejected as misleading.

The purpose of this oversimplified example is to indicate that each seer sees what he already believes is 'out there' and his seeing confirms him in his 'metaphysics'. We shall learn little concerning the essential possibilities of vision from either so long as we participate in the 'reality assumptions' which contrarily reduce the visual possibilities of the tree. And if the modern is more likely to side with the 'cartesian' that is to be considered as nothing more or less than an index of the strength of the current sedimentation of beliefs — ours seems to be a more 'cartesian' age. But even that is unsure, for the more extreme the seers of the 'clear and distinct', the more the likelihood of a revival of the seers of the 'misty and animated' may be, precisely because the 'obviousness' of a whole range of visual possibilities begins to strike such seers.

But a phenomenology of vision begins precisely with the 'possibilized' range of visual possibilities. By relativizing the 'beliefs' of both seers, the vast and flexible range of visual possibilities forms the first index for a deeper investigation of vision. The 'suspicion' of epochē which 'brackets' beliefs is balanced positively by a search for a full range and continuum of possibilities which may be overlooked within the closed contexts of the seers.

Already, in the imaginative case discussed, a second order question may be raised concerning any clear and necessary connection between objects and vision. The 'cartesian' tree is different than the 'spirtualist' tree for both are reductionist trees which belong to a different range of visual possibilities. I shall call such reduced trees *objects* in the strong sense of the word, but the 'open' tree of the expanded continuum of visual possibilities I shall call *Other*. The suggestion is that there is an essential difference between Otherness and its reduction in objectification. In relation to vision this is the difference between the first and the second reduction. The second reduction, the reduction *of* vision, is the subtle step which is taken when otherness is reduced to object, when the Other is seen as (mere) object.

The implication of this distinction in the phenomenological context is that vision, taken in its richest sense, is not what gives us the Other as object, but the deeply seated 'metaphysics' which are so much a part of our 'seeing'. A phenomenology of vision must show that more is given to vision than is allowed by such implicit beliefs and that the otherness which is experienced in vision is far richer than objectification although objectification remains a reduced possibility within the vaster visual richness which is given. The sedimented beliefs, however, 'cover over', 'hide', or make us 'forgetful' of the richness which is nevertheless present.

But the route to a recovery of the richness of Otherness as opposed to objectivity is not necessarily a direct one. Thus the second step in this 'demythologization' of the relation between vision and objectivity is one which gradually expands the sense of seeing towards its richer possibilities.

First, note that vision is not alone in giving us distance and otherness. Hearing, for example, has long been recognized as a 'distance sense'. The distant approach of the Long Island Railroad train (overdue as usual) gradually spreads its auditory 'doppler effect' to my ears and gives me the sense of auditory distance and motion which is so familiar in daily life. The echo from the dark well also presents me with an auditory depth and distance. And if I place first a marble in a box and roll it about, followed by a pair of dice, the listener almost immediately is able to identify the shapes of the objects by sound. We may soon discover that we identify objects by sound and have an amazing ability to discriminate and differentiate things auditorily. Hearing gives us 'objects' but hearing has not been so closely linked to objectivity.

Touch also can give us objects, in some ways more 'fully' than sight. I may grasp in my hand in an all-at-once feel the full roundness of the ball which visually appears only in profiles. Moreover, there is also touch at-a-distance. Merleau-Ponty points out that the blind man feels the world *at the end of his cane*, not primarily in his hand. He is extended into the world through his cane. And, although we seldom think of it in this way, we are 'normally' feeling at a distance through being 'embodied' tactilely and kinesthetically in the use of machines.⁴ The driver of an automobile, particularly if he is a sport driver, feels the surface of the road not just the resistance of the steering wheel. Yet even if hearing and touch give us 'objects' at a distance, objectification remains less closely related to our traditions concerning these senses. This should lead us to suspect that objectification may have as much or more to do with those 'traditions' than it does with the essential possibilities of the perceptions involved.

Vision has, on the other side of this question, a whole range of possibilities, 'forgotten', or 'covered over', in which the experience of a loss of distance and a positive intimacy with otherness often associated with the other senses also occurs. The example of the 'spiritualist' seer, as far fetched as it may have seemed earlier, has its counterpart in certain experiences of persons seeing for the first time. J. M. Heaton in *The Eye: Phenomenology and Psychology of Function and Disorder*, reports that patients first gaining eyesight through medical means, "learn to recognize colours before form and may even be able to name them before being able to recognize simple forms. *At first colours are not localized in space and are seen in much the same way as we smell odours*," (italics mine).⁵ The object gradually takes its shape through vision and its "first" appearance is far from that of the clear and distinct object which is only 'later' attained. Not only does this imply that we 'learn' to see and with this learning there is the constant presence of a quite structured learning according to the dominant 'metaphysics' of the milieu, but in the learning certain possibilities of vision are subtly selected and others gradually 'covered over'. The 'cartesian' preference for the geometric values shapes or forms over colors while an 'artistic' preference may have countervalue.

The examples of visual learning through which the object is attained may also be reversed. The pervasiveness of color as non-localized and coming at one like odors has often enough been reported by youthful drug users. Under the influence of hallucinagenics the habitual ways of viewing are deconstructed in a reversal of the blind man's coming to sight experience. A report of one scientist under the effects of mescaline indicates his shock:

One believes in hearing noises and seeing faces, but everything is one.... What I see, I hear; what I smell, I think. I am music, I am the lattice work. I see an idea of mine going out of me into the lattice work. This is not a metaphor, but the perception of something coming out of me ... everything was clear and absolutely certain. All criticism is nonsense in the face of experience.⁶

The examples of the blind man and the drugged scientist are extreme, but similar characteristics have often been noted in the development of visual learning in children. Koffka and Piaget, as distant as they may be, have noted the 'animistic' quality of perception in children. The world first appears as 'alive' to the small child and objects appear as repelling or attracting rather than being reduced to mere qualities such as color or shape. Objects are 'first' experienced as *in relation* to the actions of the child himself. The Teddy-bear illustrates all the above.[7] The 'later' attainment of objectification by which the object becomes a 'cartesian' object is an acquisition only through effort.

The learning to see the world as a 'cartesian' (or a 'spiritualist') is a task, but a task which is never completely successful. The essential possibilities of vision may be 'covered over' but they do not disappear. Thus even in normal and ordinary cases a phenomenology of vision uncovers the same drama made more pointed in the extreme cases above. In daily life we often overlook the animation and liveliness in which the visual 'object' is presented to us. 'Abstract' figures in cartoons are grasped with the 'naivete' of the child — I go to the cinema and a cartoon feature is presented in which there are merely dark colored dots lined up before an opening in two solid black lines. A single dot approaches to this line of oscillating dots and attempts to shove his way to the front of the line to the sounds of mumbled and unintelligible sounds. Although both sights and sounds are 'abstract' and ambiguous, I cannot help but see what the others also see, a cartoon as a line of people trying to get into a door with the queue-jumper trying to shove his way in. My vision in spite of myself does not give me a 'cartesian' object as such, but only by special acts of detachment can I approximate such an 'abstract' seeing.

In short, what I am isolating here is an essential feature of vision: *Vision is essentially situated within some set of 'beliefs' which influence what is 'taken' as vision — we cannot find a 'presuppositionless' vision — but at the same time the polymorphy of vision always exceeds the sedimentation of those 'beliefs'*. The point is in keeping with the insights of the later Merleau-Ponty. If we speak of the sedimented perception as a 'naïve' or 'pre-phenomenological' vision, we must note that vision 'allows' itself to be so taken, as in the 'cartesian' or the 'spiritualist' senses, but that as understood phenomenologically vision holds other possibilities as well in its essential *polymorphism*. In the *Visible and the Invisible*, Merleau-Ponty notes, "I say that the Renaissance perspective is a cultural fact, that perception itself is polymorphic and that if it becomes Euclidian, this is because it allows itself to be oriented by the system."[8]

Thus if we were to anticipate the outlines of a more complete phenomenology of vision it would be from the heart of this polymorphic richness that the inquiry would take its basis. The study of the actual forms of vision, as in the history of art, would indicate some of the diversity of 'systems' which have already occurred in relation to polymorphy. But at the same time this very richness is the source of a 'weakness' and 'ambiguity' within vision, the weakness which allows vision to be dominated by a 'metaphysics'.

It is in the phenomenological awareness of this ambiguity and open texture of polymorphy that its own 'return' to experience arises. The *epochē* which purposefully distances experience and its 'beliefs' and which reveals the polymorphy of vision to be 'inexact' but rich, must return to its own first insights and draw implications from this discovery or recovery. The 'demythologization' of a 'cartesian' vision which objectifies the thing is a 'demythologization' which sees objectification as *one* structured, but now relativized possibility of vision. The same applies to the possibilities of the 'spiritualist' vision. But implied in the preference for overwhelming richness of a phenomenological taste, there is the harboring of a secret position, a position which believes that other possibilities may also emerge to re-orient vision.

I would like to suggest in a somewhat speculative vein that such a change is already subtly taking place in the midst of our heretofore dominantly 'cartesian' culture and that, ironically, it is the very triumph of that culture in technology which has launched this change. Merleau-Ponty located the relations between what I have been calling a 'metaphysics' and perception in his later life as a doubled thesis: "1. There is an informing of perception by culture which enables us to say that culture is perceived . . . 2. this original layer above *nature* shows that *learning* is *In der Welt Sein*, and not at all that *In der Welt Sein* is *learning* . . ."[9] But this learning, which occurs in a change of vision, is part of phenomenology's own immersion in the world and history. The 'demythologization' of extant perceptions of culture becomes possible only by opening the way to other such perceptions.

I suggest that a possible key to such a change in contemporary culture revolves around the technology of vision, the television and the cinema, and is symptomatically located in what I shall call the *cinematographic possibility*. Its appearance occurs in simple experiences. On moving to Stony Brook, a campus still under construction on a site typical of instant universities, there abound mud fields, and an as yet incomplete steam system which belches vast clouds of steam from the earth, I find that looking out of my window at the opaque stare of the dark windows of the chemistry building next door a certain effect occurs. The profile of the hard-edge brick and concrete and glass in the midst of the chaotic surroundings suddenly appears as if I had seen this Antonioni movie before! Or, on driving to Ithaca College, crossing the Hudson, I look up and notice the quickly flashing by curved gooseneck street lamps moving across the windshield and again I feel as if I were experiencing "Easy Rider." There is a momentary inversion of feeling, not that a movie is like reality, but that reality is like the cinema, in a cross-sorting of the metaphor.

Not only do students confirm such experiences as common today, but a scanning of psychological literature even down to clinical reports discloses that dreams, memories, imaginations are described as 'movie-like' — needless to say, such an effect was not possible prior to the invention of the technology which allows one to 'see' his imaginations and dreams in such a way.

I am certainly far from the first to have noted this phenomenon. Nor am I in the vanguard of those who have begun to detect as yet difficult to isolate differences between the sensibilities of the post-television student and the rest of us. What I am suggesting is that a phenomenology of vision, centered in a concern with man-machine relations which in this case are the 'media'. may begin to detect a shift of vision. Within this set of nuanced clues there is emerging a sense of *vision as viewing*.

I am suggesting that although related to vision as an objectifying gaze, vision as viewing holds enough differences to be the harbinger of a change of vision. Return for a moment to our seers, this time with an eye to a new way of seeing. Our 'cartesian' seer is a seer who is an *observer*. His preferred world of vision is one in which, amidst all the possibilities of vision, those objects which are most 'clear and distinct' stand out in their colored, shaped, and extended forms, fixed before his gaze in an ideality which stops them for observation. Even motion, which poses a complexity for him, is stopped by breaking it up into units. Thus the ideal object of the objectifying gaze is one subjected to the analysis which gives a static (and mute) object.

But the cinema viewer whose vision moves from observation through the slight, but profound, distance to that of a *spectator* viewing a *spectacle* increasingly idealizes a world of movement, of dissolving, transforming, changing and 'distorted' *images*. Substances no longer remain stable: the woman becomes a donkey before one's eyes; the tree gradually takes the shape of a man — in short, in the cinema the world of 'magic' has again re-emerged as a 'normal' phenomenon of vision. The mist and wind of our 'spiritualist' seer is psychedelically present in living color at the local cinema and the preferred static entities of the 'cartesian' are replaced there with a new preference for dissolving and changing images of this magical world.

Now take a further speculative step which relates this everyday occurrence to the subtle, but largely implicit learning in which "culture is perceived" as Merleau-Ponty suggested. In that nexus there begins to be common the cross-sorting of the experience of the cinema with that of daily life and inversely, daily life begins to be perceived as more and more 'like' a cinema. Daily life begins to become a spectacle. That something like this is occurring in contemporary culture has already been pointed out, in spite of woefully inadequate epistemologies, by the prophets of the media (McLuhan and Ong, for example). The vision of viewing which sees the cartoon, the seven-o-clock news, the Olympics, and the man landing on the moon through the 'images' of the television makes of each of these occurrences the same near-distance of the spectacle.

The technology of vision transforms, while 'extending' the vision of him who sees. The 'objective' attributes of space and time take second place to this *near-distance* which is made possible through the tubes. 'Objective' distance becomes relative in the quasi-immediacy of the television — Vietnam is in the living room and its politics cannot be evaded. My mother, long isolated on the plains of Kansas, reports that since television the farmers all now talk politics to the relative loss of the predominance of local gossip. The same occurs with 'objective' time. In the very same evening one can see the same star either age or regain youth in a revival showing of a *retrospective*. Phenomenologists have, of course, pointed out many times that 'objective space and time' is constituted and not immediate — but the media, the cinematographic possibility makes that point a dramatic, but ordinary occurrence.

The nearness of near-distance is also matched by a 'distance'. The Other remains partly disembodied as 'image'. It is the Other I see and not the lines or the electronic oscillations on the screen, but the Other as quasidisembodied. The Other is not the Other with whom I may speak or have full communication and as I fall farther and father into vision as viewing the hunger for touching arises. But a frantic and artificially created situation for touching fails to bring the Other close. This, too, may be seen in contemporary culture.

For if the cinematographic possibility of vision as viewing becomes 'real' and if it follows the prevous patterns of sedimentation such that it takes the place of a dying 'cartesianism', then vision also becomes increasingly a matter of living as *spectator*. It is the spectator who stands in the relation of near-distance. For him the world is insubstantial, open to change at any moment. It is a kind of play or drama in which the plot may change at the very next viewing. But unlike the vision as objectifying gaze of the 'cartesian', the spectator is more involved in the world-as-image. The 'cartesian' remained disinterested and distant, but the spectator is already a quasi-partici-pant who is caught up in the drama he views. His emotions, his values, his fantasies and worries, are all involved in the spectacle.

But at the same time he is far short of the existentialist ideal of the committed self. The viewer goes from show to show and increasingly the long-term project becomes difficult if not unimaginable. The near-distance of the spectacle world has its focus in the elevation of the *immediate*, but an immediate which also remains in a kind of 'aesthetic' distance as spectacle. Thus the spectator whose vision is viewing begins to move from the sense and thrust of vision as objectifying gaze but remains short of finding in the polymorphy of vision the possibilities which would enrich vision as *Vision*.

The 'aesthetic' quality of vision as viewing only partially involve the seer and thus the otherness of the Other remains distant. So long as the appearance of the Other occurs in a setting in which human 'transcendence' crosses with objectification as a reduced possibility as in Sartre's descriptions of relations with the Other, the Other remains *object*.[10] But to transform the Other into image likewise remains short of the phenomenon of the Other in genuine otherness. Merleau-Ponty, I believe, points the direction away from the implicit objectification found in vision when he indicates that the 'look' only objectifies when it, "takes the place of possible communication."[11] I think I need not say that the *face* of the Other also *speaks*.

Within Vision there lies a deeper possibility than either vision as the objectifying gaze or as viewing. The otherness of the Other is also a *relation* which implicates me and insofar as the anguish of discovering self-objectification is the source of self-creativity, it is also the anguish of having to give up the last vestige of solipsism whose roots lie in the 'covering over' of that deeper relation. There is a need for a phenomenological restoration of Vision.

NOTES

[1] Don Ihde, 'Some Auditory Phenomena,' *Philosophy Today*, Vol. 10, No. 4/4, Winter, 1966, pp 227-235.
[2] Walter Robert Goedeke, 'Ihde's Auditory Phenomena and Descent into the Objective,' *Philosophy Today*, Fall 1971, pp. 175-180.
[3] See Jacques Derrida, L'*écriture et la différence* (Paris: Éditions du Seuil, 1967).
[4] See Don Ihde, 'A Phenomenology of ManMachine Relations,' *Work, Technology and Education*, edited by W. Feinberg and H. Rosemont (University of Illinois Press, 1975).

5 J. M. Heaton, *The Eye: Phenomenology and Psychology of Function and Disorder* (London: Tavistock Publications, 1968), p. 42.
6 *Ibid.*, p. 45. Note also a similar report in C. Castaneda, *The Teachings of Don Juan: a Yaqui way of knowledge* (New York: Ballentine Books, 1972).
7 *Ibid.*, pp. 47-8.
8 Maurice Merleau-Ponty, *The Visible and the Invisible* (Northwestern University Press, 1968), p. 212.
9 *Ibid.*, p. 212.
10 See Jean-Paul Sartre, *Being and Nothingness* (New York: Philosophical Library, 1956), pp. 361-413.
11 Maurice Merleau-Ponty, *Phenomenology of Perception* (London: Routledge and Keegan Paul, 1962), p. 361.

CHAPTER 8

BACH TO ROCK, A MUSICAL ODYSSEY

Followers of the McLuhan style of thought concerning the influence of media often point out that the invention of high fidelity recording is to music what the printing press was to writing. Just as books from a printing press became both easily reproducible and inexpensive and thus created the conditions for a wider literate culture, so inexpensive and widely available recordings create the conditions for a wider musical culture.

But *is* recording to music what printing was to writing? My answer is that in the deepest and broadest sense it is — but precisely because it is, some unexpected changes in musical culture may be expected. My thesis has its roots in both external trends and in personal experiences which have become more dramatic in recent years.

An external symptom is found in trends concerning classical music recording sales. Market analysis indicates that this class of music sales has not only remained stable in an otherwise expanding market, but that its customers are growing older. This, say the analysts, indicates that youth are not being converted to classical and 'serious' music.

At first I was inclined to think little of this problem, perhaps allowing myself the opinion that this is about what one could expect in the light of basically unimaginative public music education. Record sales probably showed the same tendency for mediocrity and a lower denominator as commercial television (also thinking myself a musical 'elitist' at heart). I now regard this prejudice as false.

It takes little reflection upon the wider musical situation with youth culture to note: (a) that it is pervaded by music to a degree seldom achieved by earlier generations. Youth is very much 'into music." Music is so pervasive, in fact, that it has become a symbol within youth culture. The mass gathering of the rock festival, the omnipresent stereo set in dormitory rooms, and even the occasion for the first riot in a National Park when oldsters and guards at Yosemite objected to an informal rock session, all center in musical events. (b) Although some might debate the absolute worth of youth music, it seems to me that much of youth culture's music is significantly superior to that of the swing-ballad era of my youth. Youth culture's music is considerably varied in style from folk to country to rock; inventive in hybridization, folk-rock, to all other conceivable combinations; culturally open, Oriental influences to revivals of strains of classical instruments such as harpsichord and lute; and is purposefully experimental in contrast to much earlier program-produced 'pop' music. (c) Moreover, youth culture's music is largely indigenous and I suspect much of the improvement comes precisely from the overthrow of the select group of formula 'professionals' who dictated the style of earlier radio and record productions. The music of youth culture has invaded the wider cultural scene. 'Adults' now play the Beatles and their brethren rather than scorning kid music.

If the above is the case, one could already point to a general parallelism between the introduction of recording and the earlier introduction of printing. The wider universalization of music, the proliferation of styles, and a democratization of the music-making process, are all analogous to what happened at the dawning of the 'Gutenberg era'.

But these general parallelisms do not indicate why, within the proliferation of styles, classical music has apparently been shunted aside or relegated to a minor role. If youth culture is musically more healthy than it was two decades ago, why is there not an increase in or even stabilization of classical music appreciators among them? The answer, in part, lies in something occuring within youth culture itself. Upon inquiry many students who are seriously interested in music will also respond that classical music is 'head music' while their music is 'body music'. Is this merely to assert a prejudice for rock over Bach? I think not, rather it has something to do with a relationship between musical experience and its concrete embodiment in reproduction.

I recognize here that by turning to personal experience and generalizing upon it I risk being speculative at best – however what I outline below has apparently also been a common experience. When I first acquired a high fidelity stereo set as a graduate student, our first record purchases were all of the 'serious' variety with special emphasis upon baroque, particularly Bach and Vivaldi. In ecstasy I would listen to the disciplined, artful, magnificence as this music filled the room and my consciousness. However, it took little time to note a flaw – the more I entered into the music the more I became aware of auditory distractions, particularly those from the set itself. Any scratch, any barely audible hum, any interference became annoying and threatened the enchantment of the experience. Even more dramatically I noted, each time I went to a concert, how much 'purer' the actual live performance was.[1]

The music demanded *purity* of sound, and reproduction, no matter how good, always remained short of this live purity of the concert hall or chamber. I am claiming here that the *context* in which the music developed historically turns out to be more important than it might at first appear.

Over the years my record library expanded and my wife and children exerted their influences and tastes as well. Music was played to fit – and sometimes to create – the mood. Many occasional evenings were spent not merely listening but dancing expressively to the stronger beats of Peter, Paul and Mary, Joan Baez, and the Victory Baptist Choir. Later, Country Joe, Cream, and Jefferson Airplane were introduced and I began to discern on the edges of my consciousness the difference between 'head music' and 'body music'.

The difference is not one of mere volume nor of strong beat – although both are elements of the musical demand of youth culture music. More holistically I would say that rock and its relatives exert a call which orders either rejection or participation. It is too noisy, too insistent to be ignored. It is 'either/or'. The music in this sense demands a 'conversion'. Its call is enticing and vibrant. Its dynamism is such that after some listening it is no wonder that Bach seems so 'tame' by comparison. Youth music is seductive within its very noise.

Nor is the difference between 'head music' and the seductiveness of 'body music' one of complexity compared to simplicity. Both have their own styles of complexity and within the respective genres examples may easily be found illustrating a range from simplicity to complexity and from good to poor quality. One must learn to listen to each if the lesson of the music is to be learned. Rock almost always appears to the beginner as 'noise'. But *within* the noise music occurs. The first time I listened to Cream, its noise and its repetition were the only factors I could discern – amplified musical paganism. Later it became domesticated and the repetitions receded to the background against more subtle modulations which filled the foreground. (I have not, nor do I intend to attain the quasi-mystical state reported by the 'true believer' which in the presence of the fully amplified piece, the loudness so near the pain threshold that a kind of 'silence' is reported – although I believe I can recognize that experience as an extension of what I already know. In this extreme state, however, the cilia of the ear are eventually damaged. If so, the 'conversion' to rock becomes permanent in a sense because the more delicate sounds of 'head music' would be forever lost to the hearer.)

There is a third factor which began to emerge in listening to the 'harder' types of 'body music' – the search for purity experienced with classical music is lacking. In fact, instrumental tonal purity is irrelevant here. The sound of rock begins electronically, reproduction and amplification are a part of its very embodiment. Thus the process of electronic reproduction does not get in the way – the music is itself an electronic creation. I am suggesting that just as the live performance of chamber or concert music is the medium, the historical and actual context in which the classical form of music developed, so the transistor and the amplifer are the context of the harder forms of youth culture's music.

The differences that emerge between 'head music' and 'body music' are easily felt, but somewhat difficult to express. Part of the problem is found in the actual use made of the difference within youth culture. As used there are two meanings which are not equivalent. The first is a deprecating meaning – 'head music' is music which is composed. It is music which youth culture believes is mechanically created according to a theory or 'science' of music, whereas most forms of youth music are thought to be more subjectively expressive and spontaneous. This semi-romantic notion of 'body music' repeats a common theme of youth culture: whatever is expressed must be genuine and personal. Here the negative valuation of 'head music' is both anti-intellectual and yet potentially creative.

The anti-intellectualism is merely naiveté. Surely many composers of one, two, and three centuries ago deeply *felt* their music in spite of composing within an accepted idiom and, contrarily, one can also point out that the nascent romanticism of subjective expressivity and spontaneity is also an idiom which has its own 'laws'. But the anti-intellectualism, even if uninformed, does have a functional role. It allows the new idiom to emerge. The movements of the history of thought, in music as well as in other areas, seem to be accompanied by iconoclastic polemics against previous forms. The polemic often is the tool (and sometimes excuse) for sharpening the edge shapes of the new tradition and form.[2]

But more profoundly and genuinely the difference between 'head music' and 'body music' is an attempt to express a difference in the felt response to the two types of music. There is a sense in which the common meaning is too gross. All music listened to seriously is some kind of 'body music' – one hears not with ears alone, but with one's body. The Bach fan, listening with closed eyes to the musical presence of an excellent fugue experiences a set of bodily tonalities which 'move inside' him. He may feel an inner sway to the beat and rhythm, a chill to certain passages, and be 'filled' with the music. He is listening with his body. But in spite of that one must characterize the felt bodily response as properly reflective of the music itself. These tonalities are restrained, measured, and as it were, controlled.

There is even a sense in which the bodily focus, though distributed throughout the body, is more often focused higher in the body itself than with rock. And this provides a clue to the emphasis within 'body music' as used within youth culture. Rock, too, is heard with the body. But the insistent demand of the music resonates, almost against any will of the listener, 'inside one'. I find that the focal location of the rock response is often lower in the body — the bass notes are felt in the chest and stomach — even though, again, the music resonates throughout the whole of one's body. At the extremes of amplification where the quasi-mystical trance of pain—sound occur, it becomes almost literal that one hears more with the body than with the ears (head).

Rock is the celebration of amplification and the electronically embodied instrument. Even the voice is 'electric' through the use of echo boxes and the deliberately created 'distortions' of rock singers when they hold the mike to their lips or wave it in the air. The *body* of 'body music' is electric.

A more precise parallel between the history and development of recording and printing now emerges. When the printing process was young, it first imitated the familiar hand produced script. Early printing used gothic letters, was elaborate, and expensive. Its stylistic paradigm was extant handscript merely transferred to the new process. Only much later did the potential of the press begin to be realized — the development of the sparse, simplified, and easily read letters of modern typescript grew out of the potential of the medium.

Contentwise a similar process followed printing. The Protestant Reformation saw in the press a means of radicalizing religious culture. Luther's "German nightingales can sing as sweetly as Latin larks" was an insight into the culture-transforming power of the dissemination of the (printed) word. Printing only gradually expanded from Bibles and theological tracts to business accounts and advertisements.

In a faster time period the same developments in recording may be seen. At first what had always been heard was re-produced, whether the sound of big bands or Mozart. Later came the demands of the medium itself. Amplification, electronic sound, becomes its own instrument. Re-production becomes production. And, not to stretch the point too far, the 'revolutionary' implications for culture within youth culturer's music are not too far disjunctive from the radical potential seen by Luther's demand for cultural change.

But in the process, just as Latin and gothic script — remaining beautiful, artistic, and to be appreciated — began to recede as the main style of the 'Gutenberg era'. Bach becomes more clearly 'period', more museumlike in the perspective of youth culture.

I am not prophesying the death of Bach. But if the printing-recording parallelism holds, it does mean that Bach will have a harder time in the future. There will remain 'two culture' people just as in the present omnipresence of television, reading continues to hold its own (unless, ominously, those members of youth culture who claim having 'transcended' literacy win the day).

Nor am I prophesying the eclipse of 'serious' music. The innovative movements of the twentieth century, from the explicit development of new electronic instruments, to computer composed music, to experimentation with atonality, to the contemporary movement within the arts to make whole fields – in the case of music the field of sound – of experience a kind of 'art', are all parts of a general movement symbolized by the 'Bach to rock' example I have developed here.

These trends and the experimentation within the 'new music' may not yet have found their proper voice. Much 'serious music' is still hampered, in the view of youth culture, by an overly metaphysical-theoretical mindset. It has not allowed the electronic instrument to reveal its own style. But just as gothic frillery in the first steel buildings disappeared, so will the new genre gradually be purified of the past so far as it is unnecessary.

What I am indicating is that within the sympiomatic shift in *musical technology* a deeper shift of sensibilities is also going on. The McLuhanites are at least partly right in labeling the era in electronic one. The 'electric' is a new – but I suspect transitional – *symbol* for the shift in sensibilities. We tend to mold our concepts of ourselves upon our concepts of the world. Our active and operative 'myths' which contain this self–world interpretation function in terms of key symbols or metaphors. In an 'electric' era we model our minds upon the electric computer. The cinema, another electric creation, has become so pervasive in the way in which we have begun to understand ourselves that even psychological literature has begun to liken our dreams, our fantasies, our visual imagination to the 'movie-like'.

The 'electric' world is a world of 'flow', its images are suggestive of transmutation, transformation, and the melting of distinction. In music, again particularly among youth, the whine and microtonic 'flow' of the sitar and the 'electronic instruments' 'infinite flexibility' embody the flow of the electric. In cinema the flow of images magically transforms our seeing so that images melt into one another and transmute the entities of the screen in such a way that the 'metaphysics' of ancient demonology and witchcraft become real within the possibilities of film.

The 'electric flow' image of much contemporary culture is in contrast to the mechanical, the clear and distinct atomism of the recent and perhaps still dominant past. The change in sensibility is symbolized by the shift from 'mechanical' to 'electric'. The world becomes 'Heraclitan' before us rather than 'Democritean'.

But the McLuhanites are wrong when they tie this shift of sensibilities to a shift from one to another dominant sensory form. They have almost convinced us that the era whose presence they see slipping into ancient history is one which is *visualist* in sensory form. For them vision is the paradigm sense which gives us the mechanical, the clear and distinct differences, the atomism of a 'Democritean' type. The electric era, they say, re-introduces an *auditory* imagery based upon flow, the melting of differences, the emphasis upon motion — the 'flow of the electric'. Television, records, radio all destatify us by returning us to a more auditorily oriented culture.

Historically they have a point. The early scientific literature did emphasis much *visual metaphor*. It did utilize measurements which were necessarily embodied in spatial—visual forms. The 'ideal observer' as the limit idea around which Newtonianism revolved was metaphorically a looker at a silent world of mechanical motion.

In all of this, however, the reduction of early modern scientific culture was *not* so much a reduction *to* vision as the McLuhanites hold, but a reduction *of* vision. What is needed is a re-evaluation of the full range of possibilities within sensory experience. In fact the metaphor of the 'electric flow' has already begun to transform our vision itself. The cinema and television have already begun to teach us that vision flows, blends, transmutes and transposes. And contrarywise, we ought to learn within musical experience that hearing can become keyed to the accurate tone, to precision, to clear and distinct differences. It was — precisely in the perfectly beautiful, if 'mechanical' sounds of Mozart and Vivaldi.

Merleau-Ponty has understood more profoundly that "perception itself is polymorphic and that if it becomes Euclidian, this is because it allows itself to be oriented by a system." (*The Visible and the Invisible,* Northwestern University Press, p. 212). If both vision and hearing today become attuned to the potentials of 'electric flow' it is because our perceptions are concretely situated within an emerging metaphor, a newly oriented system.

Beneath the shift from Bach to rock lies the more profound shift of metaphors and sensibilities. Its embodiment lies in our *technology* and its relation to polymorphic perception. If today we have just begun to hear the world in a different way it is because we have already begun to exist in the world differently than before. The intellectual task, philosophy's fundamental thinking, is to begin to make that shift more apparent and to discern its genuine as opposed to its inauthentic possibilities.

NOTES

[1] The difference between the live and the recorded performance is also affirmed by the rock listener, as Robert Messing, a student at Stony Brook, pointed out in a protesting response to this essay:

"In fact, a recorded re-presentation of a rock concert emerges as radically incongruous from the live event, and this incongruity does not exist for the classical connoisseur listening to a recording, in that the ambiance of a live rock concert is characterized by sound levels which, from a practical consideration, cannot even be approximated at home, and additionally in that the rock concert atmosphere is essentially typified by the presence of several thousand tripping, dancing, and occasionally screaming freaks who collectively attain levels of mental energy that would probably kill a typical classicist of moderate temperament."

What Messing is pointing out is that this difference between the live and the reproduced concert is immense in that the fullness of volume and presence of the live concert is absent from the record. But is the instrumental quality absent? Messing agrees that rock is the celebration of amplification — the live concert gives *more* of that!

[2] Mr. Messing, interpreting his understanding of youth culture, indicates that the difference between 'head music' and 'body music' might, "better be characterized as that between reason and passion; not that reason may not be passionate nor passion reasonable but rather that youth demands from its music, as it demands from life, a brutally seductive intensity and emotional excitement that will grab the listener up from boredom and shake him loose from his foundations in a struggle to live continually, as it were, on the verge of orgasm. Youth has no use for disciplined, artful magnificence to which one must reach out with patience. If youth doesn't listen to classical music it is because, by virtue of its beauty, it finds it boring."

DIVISION THREE

PIONEERS IN THE PHILOSOPHY OF TECHNOLOGY

CHAPTER 9

HEIDEGGER'S PHILOSOPHY OF TECHNOLOGY

Among the few philosophers to date to have taken technology seriously, it should be apparent that Martin Heidegger is a pioneer in this field. He was among the first to raise technology to a central concern for philosophy and he was among the first to see in it a genuine ontological issue. This is the case in spite of the dominant and sometimes superficial interpretations of Heidegger which see in him only a negative attitude to technology.

It will be the aim of this essay to examine some of Heidegger's main theses concerning technology and to elucidate the strategies which motivate these. To make the task managable, I have chosen to limit myself to his 1954 lecture, 'The Question Concerning Technology', and the earlier foundational work, *Being and Time* (1927).[1] And as an interpretative device, I shall read these two works retrospectively. That is, I shall isolate what emerge as the principal themes concerning technology from the lecture and then show how they reflect and are anticipated by *Being and Time*.

In so doing, I shall show how Heidegger's philosophy of technology is directly phenomenological in the sense of exhibiting the *existential foundations* of the technological enterprise. This type of phenomenology, already apparent in *Being and Time*, gives a certain priority to what I shall call the *praxical dimension* of human existence and it continues to be a key to the later work on technology.

It is my own conviction that Heidegger's philosophy of technology is one of the most penetrating to date. By examining the ontological grounds of technics, Heidegger has begun to lift technology out of its subjectivistic and merely instrumentalist interpretations and made of it a primary philosophical question. But this is not to say that this first work concerning technology is also the last word. There are implicit limitations in the Heideggerian program which lay the basis for the current misinterpretations of Heidegger and for which Heidegger himself must be blamed. I shall point to some of these along the way.

Finally, although I shall not develop this theme in substance, I do wish to point up the resultant internal need within the Heideggerian program concerning technology for the emergence of an 'aesthetic' as the counterfoil to the limitations of technology as the Heidegger sees them. The *poiesis* which characterizes much of the 'late' Heidegger's work arises directly as a response to technology. I shall point to how this is the case within the exposition.

I. THE QUESTION CONCERNING TECHNOLOGY

The question referred to by the title of the lecture is one which takes recognizable phenomenological shape quite immediately. The query is into the *essence* of technology in its *relationship* with human *existence*.

> We shall be questioning concerning *technology*, and in so doing we should like to prepare a free relationship to it. The relationship will be free if it opens our human existence to the essence of technology. When we can respond to this essence we shall be able to experience the technological within its own bounds. (T 287)

The analysis is to make the phenomenon of technology stand out such that its horizon, limit, is bared, but this in relationship to human existence. These are the typical marks of Heidegger's version of phenomenology in which the intentional arc of *human–existence relation–world* are interpreted existentially such that intentionality is best described as an *existential* intentionality.

To uncover the phenomenon, it must be freed from its layers of less adequate interpretation which, again in typical fashion, Heidegger attributes to a 'subjectivistic' understanding, here called the instrumental and anthropological definitions of technology.

> One says: Technology is a means to an end. The other says: Technology is a human activity. The two definitions of technology belong together. For to posit ends and procure and utilize the means to them is a human activity. The manufacture and utilization of equipment, tools, and machines, the manufactured and used things themselves, and the needs and ends that they serve, all belong to what technology is. The whole complex of these contrivances is technology. Technology itself is a contrivance – in Latin, and *instrumentum*.
>
> The current conception of technology, according to which it is a means and a human activity, can therefore be called the instrumental and anthropological definition of technology. (T 288)

Such a definition implies that technology is merely an invention of a 'subject' and functions as a mere, neutral instrument. The definition, Heidegger characterizes, is *correct*. But, then, in a move directly reflective of his earlier analysis of logical or propositional truth in relation to truth as disclosure, he notes that what is correct is not yet *true*.

Correctness turns out to be 'true' in a very limited sense, true with respect to some aspect or part of a larger whole. The whole, however, is more than which contains parts, it is ultimately the set of conditions of possibility which found the parts.

The correct always fixes upon something pertinent in whatever is under consideration. However, in order to be correct, this fixing by no means needs to uncover the thing in question in its essence. Only at the point where such an uncovering happens does the true come to pass. For that reason the merely correct is not yet the true. (T 289)

The phenomenological form of the argument here is that correctness is not in itself untrue, but limited or inadequate, and may be characterized as a partial truth. But the catch is that unless it is seen for precisely this it can be taken for more than a partial truth in which case it now covers over the larger or more basic truth which founds it. It then becomes *functionally* untrue by concealing its origin. Moreover, it is only by comprehending the whole which founds correctness that it can be seen as partial. Thus what is involved in taking correctness for truth is like a fallacy of taking a part for the whole. But it is also more than that in that comprehension of the whole is a necessary condition for recognizing what is a part.

Heidegger's strategy becomes clearer if it is seen that his overall theory of truth is, in effect, a complex field theory. Truth is *alēthia*, translated as 'unconcealedness', brought to presence within some opening which itself has a structure. Beings or entities thus *appear* only against, from and within a background or opening, a framework. But the opening or clearing within which they take the shapes they assume, is itself structured. Overall this structure has as an invariant feature a concealing—revealing ratio. Thus one may say that it always has some selectivity factor as an essential feature.

Understood in this way, it becomes clear that beings as such are never simply *given*: they appear or come to presence in some definite way which is dependent upon the total field of revealing in which they are situated. Preliminarily it is important to note that the field or opening in which things are 'gathered' is, in a sense, given. It is given historically as an epoch of Being. This is to say that the Heideggerian notion of truth has something like a 'civilizational given' as a *variable*. It is what is taken for granted by the humans who inhabit such a 'world'. Variables given in this sense are particular shapes of the invariant revealing—concealing structure of truth.

What is usually missed concerning this complex field theory of truth is its phenomenological role in Heidegger. The phenomenology of truth isolates the invariance of truth as the revealing—concealing structure itself, the ratio of gathered presences to what is not revealed. Thus any particular variant is but a variant upon this overall structure. Most interpreters have missed this and failed to see that Heidegger's use of the Greeks, for example, serves as a contrasting variation upon the contemporary scene in order to point up the specific features of our epoch of truth. The interpreters often miss the counterpart characterization of the Greek modes of concealedness to which Heidegger also refers.

Thus, in Heidegger's sense, one must see beyond correctness if one is to attain truth since correctness is grounded upon some framework which makes it what it is. The process by which this penetration is accomplished is familiar from *Being and Time* as well. One begins with what is called the *ontic* in *Being and Time*. But then, by what I shall call an act of *inversion*, Heidegger seeks through it an *ontological* condition. It is only through the ontic that the ontological can be understood, but the ontological dimension is in turn the field of the conditions of possibility which founds the ontic.

It is precisely this strategy which Heidegger applies to technology. The anthropological–instrumental definition of technology is functionally ontic, correct but partial, limited to a subjectivistic set of conditions. Heidegger inverts this definition by asking a question which belongs to the transcendental tradition of philosophy; what are the set of conditions of possibility which make technology possible? Technology, as Heidegger sees it, is not only ontic, but ontological.

At first such a move seems strange, but placed within the Heideggerian theory of truth, it begins to make sense in the following way. The things of technology (instruments) and the activities (of subjects) which engage them, appear as they do only against the background and founding stratum of some kind of framework. Technology in its ontological sense is not just the collection of things and activities, but a *mode of truth* or a field within which things and activities may appear as they do. Technology is thus elevated to an ontological dimension.

Technē is a mode of *alētheuein*. It reveals whatever does not bring itself forth and does not yet lie here before us, whatever can look and turn out now one way and now another. (T 295) . . . Thus what is decisive in *technē* does not lie at all in making and manipulating nor in the using of means, but rather in the revealing mentioned before. It is as revealing, and not as manufacturing, that *technē* is a bringing forth... Technology is therefore no mere means. Technology is a way of revealing. (T 294)

Technology as a mode of truth assumes the overall shape of Heidegger's truth theory. "Technology is a mode of reavealing. Technology comes to presence in the realm where revealing and unconcealment take place, where *alēthia*, truth, happens". (T 295)

A mode of truth as a variant upon the revealing–concealing invariant carries with it certain characteristics. A few of these are important to note with respect to the specific characteristics of technological truth. I have called what Heidegger sometimes calls "epochs of Being", civilizational givens. These are something like deeply held, dynamic but enduring traditions, historical but no more easily thrown over than one's own deepest character or personality. Thus for an individual it is possible to say that he stands in or stands over against that which precedes him. ". . . The coming to presence of technology gives man entry into something which, of himself, he can neither invent nor in any way make. For there is no such thing as a man who exists singly and solely on his own". (T 313) Secondly, these civilizational givens make a claim upon those who inhabit them such that some response is necessary (although variations might range from sheer rebellion to willing acceptance). And, thirdly, they have a *telos* or inherent direction which Heidegger terms a *destiny*. But, as will be noted, a destiny is not a strict determination, it is more like a direction of growth and decay. ". . . We do not mean a generic type; rather we mean the ways in which house and state hold sway, administer themselves, develop and decay". (T 312) Technology, ontologically, is what characterizes the variant of this epoch of Being, thus penetration of its essence or shape becomes a central philosophical concern if we are to understand our era and prepare a response to it. Again, technology is elevated to a seldom seem philosophical importance in Heidegger's sense.

Now every shape of truth as a variant upon the revealing–concealing ratio has a certain definiteness to it. It has an essence or structure which is not merely its genus but is the particular form of its set of possibilities which found what we take as contemporary technics. The name for this shape of technological truth, Heidegger calls *Ge-stell*. *Ge-stell* means in ordinary German, frame or apparatus, skeleton and in Heidegger's use, *enframing*.

With *Ge-stell* the essence of technology is named: "we now name that challenging claim which gathers man thither to order the self-revealing as standing-reserve: Ge-stell". (T 301) This is Heidegger's ontological definition of technology. It has the features previously mentioned of being a civilizational variant into which humans have moved; of being a mode of revealing which serves as the set of possibilities by which technology ontically appears as it does; of making a call or clàim upon humans for some necessary response; and it has a telos or destiny as a direction of development.

By introducing *Ge-stell* at this point, I have leapt over Heidegger's development in the lecture, but I have done so in order to display what may now be called his ontology of technology, the elements of which have been mentioned. What yet remains is to examine this notion in such a way that it can be seen to account for the major features of technology in its contemporary sense and to note more specifically Heidegger's claim that technology can be thought of as the *primary* mode of truth for the contemporary era. To accomplish this task I shall turn to some more specific aspects of each of the structural features of technology as Heidegger exhibits them.

Technology is a mode of revealing. Revealing is a coming to presence within a framework. Already at this level one can detect the emergent value given to *praxis* by Heidegger. In typical fashion he reverts to etymological expositions upon Greek thought which stands at the origin of our epoch of Being. *Technē*, Heidegger points out, is originally thought of as broader than 'technique' in the contemporary thought. "*Technē* is the name not only for the activities and skills of the craftsman, but also for the arts of the mind and the fine arts. *Technē* belongs to bringing forth, to *poiēsis*; it is something poetic." (T 294) *Poiēsis* is both making and bringing forth, but bringing forth is presencing and thus is a *praxical* truth. Here is already the seed for the primacy of the praxical which characterizes Heidegger's phenomenology, but at this point is is only important to see that *technē*, as with the ancients, is linked to *epistemē* as a mode of truth as bringing to presence. *Technē* reveals or brings to presence something which is possible. "What has the essence of technology to do with revealing? The answer: everything. For every bringing-forth is grounded in revealing." (T 294)

But what is revealed? Technological revealing takes its particular shape from its field of possibilities, its framework. And its framework is a particular form of the human taking up a relation to a world through some existential intentionality. There is thus some particular presumed shape to world and some particular activity which responds to that shape of the world.

The world in its technological shape, is the set of conditions which Heidegger defines as world taken as *standing reserve (Bestand)*. This is to say that the world, revealed technologically, is taken in a certain way, as a field of energy or power which can be captured and stored. "The revealing that rules in modern technology is a challenging, which puts to nature the unreasonable demand that it supply energy which can be extracted and stored as such." (T 296) This makes world a field as standing-reserve.

This view has certain consequences, for example, "The earth now reveals itself as a coal mining district, the soil as a mineral deposit," (T 296) which is to say that nature appears as a certain potential for human use. This is a *variant* upon how nature may be viewed. It stands in contrast to those civilizational variants which, for instance, regard the earth as mother and to which one does not even put a plow. Thus one may say equivalently that the technologically viewed world is a variant upon civilizational possibilities or that it is a historical *transformation* upon how nature is taken.

Heidegger argues that such an understanding of the world is a condition of the possibility for our taking up the kinds of technologies which we actually develop now. He emphasizes the transformational features of this enterprise. Thus not only is it the case that the earth may be viewed as a resource, but what was previously taken as the dominance of nature over man becomes inverted so that man dominates nature through technology. "In the context of the interlocking processes pertaining to the orderly disposition of electrical energy, even the Rhine appears to be something at our command . . . the river is dammed up into the power plant. What the river is now, namely a water-power supplier, derives from the essence of the power station." (T 297) Technology, in this sense, is both the condition of the possibility of the shape of world in the contemporary sense, and the transformation of nature itself as it is taken into technology.

Phenomenologically, for every variant noematic condition there is a corresponding noetic condition. Thus if world is viewed as standing-reserve, the basic way in which the world is perceived, there must also be a correlated human response. That, too, takes particular shape in a technological epoch. The activities of humans in response to world as standing-reserve are those of revealing that world's possibilities, characterized by Heidegger as "unlocking, transforming, storing, distributing, and switching about." (T 298) Man is taken into the process of *ordering*: "Precisely because man is challenged more originally than are the energies of nature, i.e., into the process of ordering, he never is transformed into mere standing-reserve. Since man drives technology forward, he takes part in ordering as a way of revealing." (T 299–300)

Here, once again as in *Being and Time*, there begins to emerge the primacy of praxis which characterizes Heidegger's version of phenomenology. And it is here that I shall begin to make the most specific connection with Heidegger's famous 'tool analysis' which serves as the model for his philosophy of technology. The common view of technology, related to what Heidgger calls the instrumental and anthropological view, holds that modern technology is a child of modern science. Technology is a mere tool of science or, at best, an applied science. Heidegger inverts this view and claims that modern science is essentially the child of technology. The strategy by which he seeks to show this is a reflection of the same functional inversion employed in *Being and Time*. This inversion of science and technology calls for careful examination.

There are two correlated ideas which appear at the beginning of the strategy which bear initial note. First, Heidegger grants that the contemporary dominant view of technology seeks to strongly differentiate between scientific technology and the older handwork technology. Heidegger does not deny that there are differences, but he plays these down. For instance, in granting correctness (not truth) to the instrumental view of technology, he notes that this view can bring both handwork and scientific technology under the same rubric as 'means' or as instrumental towards ends. Here the difference between technologies is merely a matter of relative complexity. (T 288–9) Secondly, the constant emphasis upon technology as *poiēsis* and as *technē*, a making in the ancient broad sense, tends to play down a difference between ancient and modern technology. But thirdly, and most profoundly, the differnce is played down strategically because the essence of technology is not itself technological, but is existential. What Heidegger does grant is that modern technology allows the secret grounds of technology as enframing to emerge more clearly, allows what was long latent and originary to be made more explicit.

Correlated with this downplay of an essential difference between ancient and modern technology is the necessary admission that modern technology is chronologically later than modern science.

Chronologically speaking, modern physical science begins in the seventeenth century. In contrast, machine-power technology develops only in the second half of the eighteenth century. But modern technology, which for chronological reckoning is the later, is, from the point of view of the essence holding sway within it, historically earlier. (T 304)

(Here one must recall the difference between history as a destiny and historiology as a chronicle developed in *Being and Time*.) The essence of technology is not chronologically prior, but it is historically, ontologically, prior to modern science itself. It is from this inversion that Heidegger makes his claim that the technological epoch is what characterizes the contemporary era. The claim is clearly reflective of his earlier explicit claims regarding the primacy of the praxical.

In the lecture the inversion first takes explicit shape regarding science and its instruments.

It is said that modern technology is something incomparably different from all earlier technologies because it is based on modern physics as an exact science. Meanwhile we have come to understand more clearly *that the reverse holds true as well*: modern physics, as experimental, is *dependent* upon technical apparatus and upon progress in the building of apparatus. (T295–6; italics mine).

This is to say that modern science is *embodied* technologically. One might very well say that one basic difference between modern science and its ancient counterpart is precisely its increasingly technological embodiment in instruments.

But if science is embodied in instruments as a necessary condition for its investigations, this is not yet to say that technology is its origin. Yet that is the claim Heidegger ultimately makes. The form the argument takes is essentially that it is first necessary to view nature as a storehouse or standing-reserve towards which man's ordering behavior can be directed. This provides the condition of the possibility for a calculative modern science.

Modern science's way of representing pursues and entraps nature as a calculable coherence of forces. Modern physics is not experimental physics because it applies apparatus to the questioning of nature. The reverse is true. Because physics, indeed already as pure theory, sets nature up to exhibit itself as a coherence of forces calculable in advance, it orders its experiments precisely for the purpose of asking whether and how nature reports itself when set up in this way. (T 303)

Thus, hidden behind modern physics is the spirit of technology, technology in its ontological sense as world-taken-as-standing-reserve. Its firstness, however, only gradually becomes clear. Such conditions are not necessarily first known, they only gradually come clear. Historiologically, then, modern science does play a role. It begins to announce what lies behind science as technology comes to presence.

The modern physical theory of nature prepares the way not simply for technology but for the essence of modern technology. For such gathering-together, which challenges man to reveal by way of ordering, already holds sway in physics. But in it that gathering does not yet come expressly to the fore. Modern physics is the herald of enframing, a herald whose origin is still unknown. (T 303)

But the origin does gradually become clear, the origin which is technology as ontologically interpreted. "All coming to presence, not only modern technology, keeps itself everywhere concealed until the last. Nevertheless, it remains with respect to its holding sway, that which precedes all: the earliest." (T 303)

Technology as enframing, *Ge-stell*, as originally, is the condition of the possibility of modern science. In Heidegger's terms this is the primacy of technology.

Because the essence of modern technology lies in enframing, modern technology must employ exact physical science. Through its so doing the deceptive illusion arises that modern technology is applied physical science. This illusion can maintain itself only so long as neither the essential origin of modern science nor indeed the essence of modern technology is adequately found out through questioning. (T 304–5)

Here the inversion is complete; technology is the source of science, technology as enframing is the origin of the scientific view of the world as standing-reserve.

Enframing is both the condition of the possibility of modern science and the field of possibilities within which it moves. Enframing is the *ontological horizon* of modern science such that what occurs within it appears as it does through its types of ordering. Such is the shape of the contemporary variant so far as world is concerned.

For the limited purposes here I shall consider that the exposition of technology as grounded in enframing, the world which appears as a standing-reserve, completes the *noematic* analysis of the phenomenological program. World is that which both stands before humans and that into which they are "thrown" in Heidegger's earlier language. Thus they must necessarily enter into some kind of relationship with this world. In the context of the contemporary era the dominant mode of revealing of world is technological, thus the *noetic* analysis would have as its task the unfolding of the range of possible responses to the essence of technology as enframing.

I have already noted that the normative response is what is called by Heidegger, the *ordering* of the world (unlocking, transforming, storing, etc.). On the surface it then appears that the human response to the world seen as enframed is the activity of calculatively ordering the disposition of resources. Thus just as nature appears, within enframing, as standing-reserve, so the human task appears as a kind of command of nature through technological means.

What is normative, however, is merely symptomatic of the essence of technology as enframing. It is indicative of the core or central destiny (telos) of world under the guise of technology. It is with the notion of destiny that Heidegger undertakes what must be considered the noetic analysis of enframing. Here, again, the standard moves of a phenomenological program appear:

(1) The noematic (world) correlate appears and is defined or described essentially. "The essence of modern technology shows itself in what we call enframing." (T 305) "It is the way in which the real reveals itself as standing-reserve." (T 305) (2) Then the question of a relationship to this essence is taken up, the noetic correlate. "we are questioning concerning technology in order to bring to light our relationship to its essence." (T 305) (3) This relationship is characterized by Heidegger as a mode of "being-in" as follows:

Enframing is the gathering together which belongs to that setting-upon which challenges man and puts him in position to reveal the real, in the mode of ordering, as standing-reserve. As the one who is challenged forth in this way, man *stands within* the essential realm of enframing. (T 305) (italics mine)

In short, the response or relationship of man to the essence of technology will be in terms of the way enframing appears. And this selectivity of a way of seeing the world contains a direction or destiny. "We shall call the sending that gathers, that first starts man upon a way of revealing, *destining*." (T305–6)

Destining, in Heidegger's terms, is not described as a determination. It is rather a telos, a direction, which at best may be said to set a framework and provide a set of conditions as an inclination. "But that destining is never a fate that compels. For man becomes truly free only insofar as he belongs to the realm of destining and so becomes one who listens, though not one who simply obeys." (T 306)

It is at this juncture that Heidegger makes a strategic phenomenological move. To recognize and identify the essence of technology, to comprehend it, is to have located it or to take note of it *as bounded*, as having a horizon. Thus by the same move that grasps technology in its essence, the possibility of becoming free occurs.

But when we consider the essence of technology we experience enframing as a destining of revealing. In this way we are already sojourning within the open space of destining, a destining that in no way confines us to a stultifying compulsion to push on blindly with technology, or, what comes to the same, to rebel helplessly against it and curse it as a work of the devil. Quite to the contrary, when we once open ourselves expressly to the *essence* of technology, we find ourselves unexpectedly taken into a freeing claim. (T 307)

What this amounts to, in the Heideggerian program, is to have recognized that the relationship to technology is not technological, but is an existential relationship and hence circumscribed by all the features which characterize existentiality. And to characterize the human response to technology, now located and limited, is to recognize that technology is (a) not neutral, "We are delivered over to it in the worst possible way when we regard it as something neutral; for this conception of it, to which today we are particularly like to do homage, makes us utterly blind to the essence of technology," (T 288); (b) is ambiguous, "the essence of technology is in a lofty sense ambiguous," "(T 314), and (c) is mysterious, "technology is not demonic, but its essence is mysterious." (T 309) But all of these are characterizations of existential intentionality with respect to the truth structure of concealing–revealing.

I have indicated that Heidegger's theory of truth is a *complex* field theory. It is complex because the structure of revealing is inextricably bound to concealing — indeed, bounded by concealing which is its horizon. "All revealing belongs within a harboring and a concealing. But that which frees — the mystery — is concealed and always concealing itself... Freedom is that which conceals in a way that opens to light... Freedom is the realm of the destining that at any given time starts a revealing on its way." (T 306) I shall not here go into the complexity of the ratio of concealing to revealing which marks Heidegger's theory of truth, but it is important to note its result for a human relationship to the essence of technology.

Heidegger characterizes a range of possible responses to technology. These range from blind obedience to equally blind rebellion. But he also allows for a free (authentic?) relationship which faces technology in its essence. But because there is such a range, there is also danger: "Placed between these possibilities, man is endangered by destining. The destining of revealing is as such, in every one of its modes, and therefore necessarily, danger." (T 307)

But what is the danger? The answer is essentially the same as the previously noted danger of taking correctness as truth, the danger of taking a part for the whole. "In whatever way the destining of revealing may hold sway, the unconcealment in which everything that is shows itself at any given time harbors the danger that man may *misconstrue the unconcealed and misinterpret it*." (T 307, italics mine)

A misinterpretation, for Heidegger, contains elements which reflect the errors possible in taking correctness for truth. They revolve around his version of mistaking the part for the founding whole. Thus, unless it is recognized that technological revealing is also a concealing (and it is from concealing that the origin of freedom arises), it can be mistaken for the totality. Technology, by its very status as a mode of revealing, may harbor this temptation.

The coming to presence of technology threatens revealing, threatens it with the possibility that *all* revealing will be consumed in ordering and that everything will present itself *only* in the unconcealedness of standing-reserve. (T 315, italics mine)

Noematically, this is the implicit claim of ultimate truth, world must appear *totally* or ultimately as standing-reserve. Noetically, the same index for **danger** can occur. By reflexively taking account of their place within world, humans face the danger that they can also be taken as standing-reserve.

When destining **reigns** in the mode of enframing, it is the supreme danger. This danger attests itself to us in two ways. As soon as what is unconcealed no longer concerns man even as object, but exclusively as standing-reserve, then he comes to the very brink of a precipitous fall, that is, he comes to the point where he himself will have to be taken as standing-reserve. (T 308)

If world becomes totally perceived as standing-reserve, then reflexively, humanity itself may come to perceive itself as the same.

In one respect this is to note that the technological mode of truth is "reductionistic." But it is reductionistic in a special Heideggerian sense because it is not that something can be 'added to' this mode of revealing which will correct it — although it appears that Heidegger himself opts for something like this alternative as a solution. Rather one mode of revealing can only be changed by in effect being replaced. Its 'reductionism' is a reductionism of disregarding the concealed, the horizon of all unconcealedness or revealing.

Although I am not particularly concerned here with Heidegger's response to the danger of technology, but rather concerned with its explanatory scope, it is perhaps well to note that his response was never well formed. In the technology essay the response was, in fact, a form of remedy for 'reductionism'. It contained two primary steps. The first remains continuous with what may be called 'phenomenological therapy'. This therapy is to address the critical question to technology, as to any truth claim, and seek to limit its hubris towards totality. Critical questioning, in Heidegger's sense, calls us back to noting the structure of the invariant *concealing-unconcealing* which limits every totality. This is the perennial philosophical task:

Because the essence of technology is nothing technological, essential reflection upon technology and decisive confrontation with it must happen in a realm that is, on the one hand, akin to the essence of technology, and, on the other, fundamentally different from it... For questioning is the piety of thought. (T 317)

The other dimension of Heidegger's response is or may be seen as an attempt to broaden and enrich technological revealing. And the enrichment, he sees, comes from a similar activity which is in its own right praxical and poetic; the enrichment is to come through a basic revival of *technē as art*. This move is familiar throughout the Heideggerian emphasis upon the primal thinking of the poet, but it is rarely appreciated as the similar-dissimilar counterpart of *technē* as technological. Art *is* technological as *technē*, but its mode of revealing opens new ways of "saying Being" as Heidegger puts it, thus is fundamentally different from *technē* as technology.

What was art — perhaps only for that brief but magnificent age? Why did art bear the modest name *technē*? Because it was a revealing that brought forth and made present, and therefore belonged within *poiēsis*. It was finally that revealing which holds complete sway in all the fine arts, in poetry, and in everything poetical that obtained *poiēsis* as its proper name... poetically dwells man upon this earth. (T 316)

Technology and art belong to the danger and possible salvation of the same epoch of Being.

II. BEING AND TIME

I now turn to a brief examination of the famous 'tool analysis' of *Being and Time* as the full anticipation of the themes concerning technology in the lecture. It is first important to note the context and role that the 'tool analysis' plays in the overall Heideggerian strategy. The analysis occurs as the vehicle by which the worldhood of the world is to be made phenomenologically apparent. World, Heidegger contends, does not just appear and neither can it be accounted for by adding up and classifying the entities within it. Such a strategy always already contains a hidden interpretation and is thus ontologically naive. Heidegger's counter-strategy is to attempt to locate what is ontological *through* a phenomenological analysis of what first appears as ontical.

This strategy appears clearly at the end of the 'tool analysis' where the ontological relationship with world appears *through* the ready-to-hand:

As the Being of something ready-to-hand, an involvement is itself discovered only on the basis of the prior discovery of a totality of involvements... In this totality of involvements which has been discovered beforehand, there lurks an ontological relationship with the world. (BT 118)

Here is already a glimpse of Heidegger's assertion of the way a whole or totality, although hidden and latent, *precedes* any individual or part as the condition of possibility for the part to appear as it does while the part, in turn, is the proximal means by which the totality itself is discovered. The ontological is discovered (literally, dis-covered) through the ontic.

Ultimately, what is ontological, however, must also be noted. Heidegger's ontology is thoroughly phenomenological, although phenomenological in the specific existential sense which Heidegger gives to the intentional arc. A phenomenological ontology is one which correlates in a unified concept three distinguishing notions. These appear in Heidegger as:

Dasein-being in-World

They are clear adaptations from the Husserlian notion of intentionality which 'consciousness' is always *of* something to which the act of consciousness refers. The intentional arc in Husserl is thus: Ego-cognizing-World. It should be noted preliminarily that the interpretation in the Husserlain context is one which dominantly sticks to a more traditional perceptual and cognitional characterization of the arc as 'mental'.

Functionally, the intentional arc remains operative in *Being and Time* but it is no longer interpreted cognitionally — it is rather existentialized such that what turns out to be basic or primary is what I shall call the *praxical*. But it remains important to recognize that the ultimate structure of Heidegger's ontology is the arc: *Dasein-being in-World*.

Heidegger's transformation of intentionality into a praxical base may be seen in two complementary ways. It is, on the one hand, a deepening of the understanding of intentionality. It is to have noted that *all* so-called 'conscious' activities are equally intentional, including such phenomena as moods and emotion and, what is more, bodily movement, such that the human being as a totality is 'being-in' an environment or world. It is true that Husserl recognized this, but he continued to interpret intentionality as if it were 'mental' instead of existential. Heidegger's tactic is one of simply cutting through the traditional mentalistic language and speaking of human existence as correlated with a world. But the second way in which Heidegger's transformation of Husserlian phenomenology may be seen is by way of seeing it as an *inversion* of Husserlian priorities. Again, Husserl already saw that the phenomenological aim undercut much theory and aimed at what became known in the literature as the 'pre-theoretical' stratum of phenomena. Heidegger not only absorbs this notion, he inverts it in *Being and Time* such that a praxical engagement with entities becomes primary over the assumed theoretical-cognitive engagement which actually characterizes all Husserl's descriptions.

This 'anti-Husserl' theme in *Being and Time* is not unfamiliar, but in this context the inversion concerning praxis and theory may be seen as the anticipation of the later explicit theme which makes technology the origin of science. I shall put the exposition in the context of its proper phenomenological strategy.

Heidegger wishes to penetrate the stratum of latent, hidden, but familiar relations with the world which characterize what he calls *everydayness*. Such a stratum constitutes, according to Heidegger, the base and limits within which subsequent specifications may be made. "The theme of our analytic is to be Being-in-the-world, and accordingly the very world itself; and these are to be considered within the horizon of average everydayness – the kind of Being which is *closest* to Dasein." (BT 94)

As already noted, the analysis takes place first in its *noematic* or world-correlate step which seeks to uncover the 'worldhood of the world'. The everyday world is the experienced environment (world-as-environment). It is through the familiar, but hidden environment that clues to the World as such are to be found.

When Heidegger turns, then, to a phenomenological analysis of this everyday environment, he argues that what is first or primary is the *praxical*. We have dealings first with things which we put to use.

> The kind of dealing which is closest to us is . . . not a bare perceptual cognition, but rather that kind of concern which manipulates things and puts them to use; and this has its own kind of 'knowledge . . . Such entities are not thereby objects for knowing the 'world' theoretically they are simply what gets used, what gets produced, and so forth. (BT 95)

Heidegger argues that to take "things" interpreted as bare entities with properties, is already to have presupposed an ontology prior to the actual investigation of human engagement with the environment.

It is from this argument that Heidegger constructs two different ways of relating to entities with the environment. These two ways of relating are well known as the distinction between the 'ready-to-hand' (*Zuhandenheit*) and the 'present-at-hand' (*Vorhandenheit*). It must be noted that both are qualitatively different relations to entities within the environment.

Heidegger's inversion of Husserl is one which makes a strong contrast between the 'present-at-hand' relation and the 'ready-to-hand' relation. The first is one in which entities (beings) appear as 'just there' and as having certain qualities or predicates. They are 'theoretically determined'. Contrarily, the 'ready-to-hand' belongs to the stratum of productive use or other forms of active engagement which characterize *praxis*. And Heidegger's strategy in *Being and Time* is to show that these are not merely two alternate modes of relation, but that one is founded upon the other, in this case the 'present-at-hand' upon the 'ready-to-hand'. This is, in effect, and action theory of ontology.

Interestingly, what prevents the contemporary era from seeing the primacy of praxis, Heidegger contends, may be laid to the door of Greek philosophy. "The Greeks had an appropriate term of 'Things': *pragmata* – that is to say, that which one has to do with one's concernful dealings (*praxis*). But ontologically, the specifically 'pragmatic' character of the *pragmata* is just what the Greeks left in obscurity; they thought of these 'proximally' as 'mere Things.' " (BT 96–7)

I have already noted in the first section of this essay, that for Heidegger whatever appears, does so in terms of a whole. The same occurs with respect to 'tools' which are what most interpreters of Heidegger on this point seem to think Heidegger is talking about. Phenomenologically, however, it should be noted that the analysis Heidegger undertakes is effectively a relational analysis in which the distinguishing features of the intentional arc are what are being described. Thus one will find that the 'tool analysis' begins with the noematic correlate, the context and entity as it occurs phenomenologically. Later and reflexively referred back to its noetic correlate, the mode of engagement which is entered into by the human exister, Dasein. The totality, Dasein-relating in the mode of the ready-to-hand, with entities within the world determines what shows itself overall.

The noematic description of the analysis begins typically with the phenomenological observation that no entity (whether in the mode of ready-to-hand or present-at-hand for that matter) occurs except in a context and against a background. Thus a 'tool' shows itself *only* as already in a context, an equipmental context.

> Taken strictly, there 'is' no such thing as *an* equipment. To the Being of any equipment there always belongs a totality of equipment, in which it can be this equipment that it is. Equipment is essentially 'something-in-order-to'. . . . A totality of equipment is constituted by various ways of the 'in-order-to,' such as serviceability, conduciveness, usability, manipulability. (BT 97)

This context in which equipment occurs has, moreover, a variable structure. "In the 'in-order-to' as a structure there lies an *assignment* or *reference* of something to something." (BT 97) This is to say that any given piece of equipment is what it is in an equipmental context and that it appears in such and such a way relative to that context. The homely illustrations Heidegger employs (ink pens belonging to the context of the desk, writing paper, etc.) show both the way in which an individual 'tool' belongs to a context and how the context is variable. But it is noteworthy that even at this first level, the whole is what determines the part. "Out of this the 'arrangement' emerges, and it is in this that any 'individual' item of equipment shows itself. *Before* it does so, a totality of equipment has already been discovered." (BT 98). Here is the model of how world is "already discovered' in hidden and latent form through the use of a piece of equipment.

What emerges from this analysis is a description of equipmentally intentional structures which Heidegger calls the ready-to-hand. It is the equipmental (noematic) context which is the condition for the manifestation of a "tool" as ready-to-hand.

> The kind of Being which equipment possesses — in which it manifests itself in its own right — we call 'readiness-to-hand'. Only because equipment has *this* 'Being-in-itself' and does not merely occur, is it manipulable in the broadest sense and at our disposal. (BT 98)

What is more, it is from this structure, that Heidegger contends one can detect a kind of *praxical knowledge* which is distinct from what we ordinarily think of as theoretical knowledge. A simply predicative knowledge of things described by properties misses this stratum. "no matter how sharply we just *look* at the 'outward appearance' of Things in whatever form this takes, we cannot discover anything ready-to-hand." (BT 98)

Contrarily, it is only in use that the distinctive characteristics of the ready-to-hand emerge. "When we deal with them by using them and manipulating them, this activity is not a blind one; it has its own kind of sight, by which our manipulation is guided and from which it acquires its specific Thingly character." (BT 98) Here the turn is made to the noetic correlate. The sight which emerges in active use, noetically, is also a field characteristic of human engagement, *circumspection*. "The sight with which they thus accomodate themselves is *circumspection*. . . action has *its own* kind of sight." (BT 98–9)

Heidegger sets off in strongest terms the difference between this praxical sight and a theoretical observation. The latter would focus its gaze *upon* the 'tool' and thus make of it an object having such and such properties – but this precisely hides the distinctive character of the entity in use. It is the peculiar manifestation of the tool in use which is the secret to praxical sight. The tool in use appears, not as an object to be seen, but recedes or withdraws.

The peculiarity of what is proximally ready-to-hand is that, in its readiness-to-hand, it must, as it were, withdraw in order to be ready-to-hand quite authentically. That with which our everyday dealings proximally dwell is not the tools themselves. On the contrary, that with which we concern ourselves primarily is the work. (BT 99)

Here is an essential insight concerning the ready-to-hand. The entity in praxical use 'withdraws' or is taken into a manifestation which is partially 'transparent'. This is one reason why the ready-to-hand may be so easily overlooked and also a reason for the inappropriateness of a predicate analysis. But it is, however, a phenomenologically positive feature of the appearance. It is, moreover, thoroughly in keeping with the intentionality analysis being presupposed by Heidegger. The human user refers *through* the tool-equipment towards world in which the work or result appears. A Thing in the mode of ready-to-hand is radically different from a Thing in the mode of being 'just there' or present-at-hand.

Although a full characterization of the mode of the present-at-hand is not called for in this essay, its relationship with the mode of the ready-to-hand is. It might be thought that the two modes could merely be variants upon concern with the world, but this is not the use to which Heidegger puts his distinction. Rather, he argues that one is the condition for the other, that readiness-to-hand *precedes* presence-at-hand and it is this argument which is both the inversion of Husserlian phenomenology and the source of what later becomes the primacy of technology in relation to science.

The themes which arise in this argument are precisely those which arise concerning technology in the later lecture. First, readiness-to-hand is a mode of disclosure. It is through the ready-to-hand that the environment appears as a 'world'. Praxis discovers Nature through the ready-to-hand. Heidegger's analysis traces this discovery not merely from a subject, but intersubjectively and on through wider and wider reaches until Nature is seen in a certain way:

Any work with which one concerns oneself is ready-to-hand not only in the domestic world of the workshop but also in the *public* world. Along with the public world, the *environing Nature* is discovered and is accessible to everyone. In roads, streets, bridges, buildings, our concern discovers Nature as having some definite direction. A covered railway platform takes account of bad weather; an installation for public lighting takes account of the darkness... In a clock account is taken of some definite constellation in the world-system... When we make use of the clock-equipment, which is proximally and inconspicuously ready-to-hand, the environing Nature is ready-to-hand along with it. (BT 100–1)

Here one sees the anticipation in *Being and Time* of the way in which the founding totality is seen through a mode of disclosure. The ready-to-hand discovers world, but only implicitly because world lies 'behind' the partial withdrawal of the equipment in its use.

Our concernful absorption in whatever work-world lies closest to us, has a function of discovering; and it is essential to this function that, depending upon the way in which we are absorbed, those entities within-the-world which are brought along in the work and with it . . . remain discoverable in varying degrees of explicitness and with a varying circumspective penetration. (BT 101)

Second, what is ultimately revealed is the world as a whole. "The context of equipment is lit up, not as something never seen before, but as a totality constantly sighted beforehand in circumspection. With this totality, however, the world announces itself." (BT 105)

Third, once disclosed, world is seen to be that in which Dasein already was, that in which Dasein has its relation of being-in.

"The world is therefore something 'wherein' Dasein as an entity already *was*, and if in any manner it explicitly comes away from anything, it can never do more than come back to the world. Being-in-the-world, according to our Interpretation hitherto, amounts to a non-thematic circumspection absorption in references or assignments constitutive for the readiness-to-hand of a totality of equipment." (BT 106–7)

Each of these themes is isomorphic with their later reiteration under the aegis of technology as the current world epoch. The connection with technology has been anticipated in the primacy of the ready-to-hand announced in *Being and Time*. Moreover, the connection between ready-to-hand and world occurs by use of Heidegger's inversion which takes a specific, but peculiar turn in *Being and Time*.

What is peculiar about the mode of the ready-to-hand is precisely the way in which the entities, equipment, manifest themselves by paradoxically withdrawing in use. This partial transparency in use functions to conceal the very context in which the equipment occurs. In noting this, Heidegger is considerably subtle in his phenomenological tactics, but, simultaneously, he begins to employ what I shall call the *negative turn* to isolate the structural characteristic he is interested in displaying.

Equipment in use appears as partially transparent, as hidden from *direct* observation. To show this, Heidegger inverts the situation and contends that the equipmental context (which is the first index for world) appears through *negativity* when the equipment somehow *fails* to function.

There are two reasons for this negative turn. The first is tactical with respect to presence-at-hand. Heidegger argues that the mode of relationship which is theoretical, the present-at-hand, *cannot* discover either equipment or an equipmental context. One does not uncover the praxical at all by adding predicates to an object. A 'tool' is not a bare physical entity to which one may add 'values', nor is its serviceability or usability seen by a bare perceptual cognition. Thus the negative turn functions, in part, to short-circuit the temptation to give an account of the ready-to-hand in terms of a theoretical metaphysics. Regarding equipment, "we discover its unusability, however, not by looking at it and establishing its properties, but rather by the circumspection of the dealings in which we use it." (BT 102)

The second reason functions as a positive phenomenological tactic by making what must be described as the partial transparency of equipment in use appear *indirectly*. Thus by this variation — no different in function than a Husserlian fantasy variation — Heidegger displays this feature of the ready-to-hand by noting that a piece of equipment which malfunctions, is unusable or even missing serves to indirectly light up its genuine function. But in the process, the negative appearance must be characterized in partial thing-like terms: conspicuousness, obtrusiveness, obstinacy. "When its unusability is thus discovered, equipment becomes conspicuous. This *conspicuousness* presents the ready-to-hand equipment as in a certain un-readiness-to-hand... When we notice what is un-ready-to-hand, that which is ready-to-hand enters the mode of *obtrusiveness*." (BT 102–3)

This is to say, that a malfunctioning piece of equipment emerges from its functional transparency and becomes a 'thing' which just lies there. Indeed, it is from this *negative* characterization that Heidegger derives the origin of the present-at-hand!

> Anything which is un-ready-to-hand in this way is disturbing to us, and enables us to see the *obstinancy* of that with which we must concern ourselves in the first instance before we do anything else. With this obstinancy, the presence-at-hand of the ready-to-hand makes itself known in a new way as the Being of that which still lies before us and calls for our attending to it. (BT 103–4)

Presence-at-hand is, in this way, dependent upon the primacy of the ready-to-hand. "The modes of conspicuousness, obtrusiveness, and obstinancy all have **the function** of bringing to the fore the characteristic of presence-at-hand in what is ready-to-hand." (BT 104)

Now, once emergent *from* the ready-to-hand, the mode of presence-at-hand can attain its own relative autonomy. It becomes possible to attend to things predicatively, theoretically. But at the same time presence-at-hand has been derived from its praxical base. This derivative character of the present at-hand carries with it, at first, the interpretation which casts it negatively as a *deficient* mode of concern. "It [equipment] reveals itself as something just present-at-hand and no more, which cannot be budged without the thing that is missing. The helpless way in which we stand before it is a deficient mode of concern, and as such it uncovers the Being-just-present-at-hand-and-no-more of something ready-to-hand." (BT 103)

I take it that this inversion is strongly indicative of both the primacy of technology and of praxis in Heidegger's later phenomenology, but it is also penultimate with respect to the ultimate strategic use to which the negative turn is put. The purpose of the analysis is to get at the world which belongs to the ready-to-hand and the inversion is but one step along that way. What equipmental negativity ultimately reveals is the latent context to which it belongs, the 'world' inhabited by concern.

> When an assignment has been disturbed – when something is unusable for some purpose – then the assignment becomes explicit... When an assignment to some particular 'towards this' has been thus circumspectly aroused, we catch sight of the 'towards this' itself, and along with it everything connected with the work – the whole workshop – as that wherein concern always dwells. (BT 105)

It may now be seen that the basic strategic and functional elements which characterize the philosophy of technology found in 'The Question Concerning Technology' were present in the much earlier opus, *Being and Time*, although they are not specifically identified with technology as such there. Nevertheless, *praxis* in *Being and Time* functions as the basic existential stratum *through* which world is revealed and as the basic realm of action *from* which sciences may arise (as processes of theoretically developing presence-at-hand).

The emphasis upon praxis as existentially basic is what characterizes the Heideggerian inversion of Husserlian phenomenology. Thus it may be said with more than a touch of correct-parallelism that Heidegger is to Husserl what Marx was to Hegel.

III. TECHNOLOGY AS EMERGENT THEME

In this retrospective reading of Heidegger on technology, I have admittedly stressed those elements which are isomorphic between *Being and Time* and the technology lecture. These isomorphisms are basic to his philosophy of technology, but there are two related anomolies concerning praxis and technology in the early as compared to the later Heidegger and it is from these that I shall lay the groundwork for the concluding section of this essay.

Being and Time does not specifically raise the question of technology, although it may easily be seen that the praxical dimension of the ready-to-hand could become interpreted as the condition of the possibility for technology. What is missing in an explicit sense in *Being and Time* is the specific characterization of world taken as standing-reserve. There is a hint of this, to be sure, in that Nature becomes *available* to the ready-to-hand. "So in the environment certain entities become accessible which are always ready-to-hand, but which, in themselves, do not need to be produced. Hammer, tongs, and needle, refer in themselves to steel, iron, metal, mineral, wood, in that they consist of these. In equipment that is used, 'Nature' is discovered along with it by that use — the 'Nature' we find in natural products." (BT 100). Here we have an anticipation of the idea of standing-reserve in a particular interpretation of Nature which is linked to readiness-to-hand.

Heidegger contrasts this concept of Nature with that which he finds in science, which in *Being and Time* is essentially an abstract nature derived from a theoretical interpretaion of the present-at-hand. In *Being and Time* he makes the *contrast* between the Nature of the ready-to-hand and the Nature of the present-at-hand as strong as possible. The Nature of the ready-to-hand "is not to be understood as that which is *just* present-at-hand, nor as a *power of Nature*." (BT 100, first italic mine). The Nature of the ready-to-hand does anticipate the notion of standing-reserve, "the wood is a forest of timber, the mountain a quarry of rock; the river is water-power, the wind is wind 'in the sails'. As the 'environment' is discovered, the 'Nature' thus discovered is encountered too." (BT 100) But this ready-to-hand Nature contrasts with the *"just there"* Nature of the present-at-hand. "If its kind of Being as ready-to-hand is disregarded, this 'Nature' itself can be discovered and defined simply in its pure presence-at-hand . . . when this happens, the Nature which 'stirs and strives', which assails us and enthralls us as landscape, remains hidden. The botanist's plants are not the flowers of the hedgerow; the 'source' which the geographer establishes for a river is not the 'springhead in the dale.' " (BT 100)

What Heidegger has not yet discovered in *Being and Time* is the profound link between contemporary science and technology. The 'science' of *Being and Time* is essentially a metaphysical and even contemplative science. It is a science derived from what may now be seen to be the ancient Greek ideal of speculation and deduction. It is not yet the science which is necessarily *embodied* in instrumentation; nor is it the science which is in the service of technology as calculative standing-reserve of the lecture.

Thus the latent technics of *Being and Time* remain either innocuous or even positive. The 'tool analysis' has often been noted to be highly selective in one respect. Heidegger chooses as examples equipment which is used 'in hand', technologies which are directly employed in work projects, technologies which extend human capacities often in terms of *handiwork*. This selectivity, I shall note, colors the entire analysis and is one element of a certain Heideggerian inadequacy of interpretation regarding technics. But first, in this context, this selectivity gives a certain tone of positivity to the ready-to-hand which is lacking in the contrasting 'abstractness' of the present-at-hand.

If the first contrast between the lecture and *Being and Time* revolves around the notion of standing-reserve as the essence of technology, a second anomaly revolves around what may be called the 'disappearance of the object' which functions differently in the early and later publications. In a sense, the *object* is what appears or is constituted *by* metaphysically based science in *Being and Time*. That which just stands there and which can be made the theme for presence-at-hand is the object. The object, which is characterized by predicates, is the noema of science in the view of *Being and Time*. Contrarily, equipment in use withdraws and is neither objectified nor does it appear as directly present at all.

The negative tone which permeates the Heideggerian analysis of presence-at-hand, however, is directed at its reductionism. The object is 'abstract', reduced, not the full and rich Thing (the springhead in the dale) which is experienced in daily life. The object is the reduced noema of scientific contemplation. It is derived from and set aside from the full existentiality of *praxis*.

By the time of the technology lecture, however, the object has also disappeared from science. Under the concept of the standing-reserve, "whatever stands by in the sense of standing-reserve no longer stands over against us as object." (T298) Here objects *and equipment* are, in effect, absorbed into the new totality. "Then in terms of the standing-reserve, the machine is completely unautonomous, for it has its standing only from the ordering of the orderable." (T 298–9) Nature, already noted as taken into technology as standing-reserve, is now accompanied by 'tools' as well. The technological world is one in which the noematic correlate is simply standing-reserve and the noetically normative response is that of ordering this reserve.

Now many critics of Heidegger see in these moves simply a Heideggerian preference for the Romantic themes of much past German philosophy — and it must be admitted that Heidegger is not blameless for offering occasion for such criticism. The implicit problem of *Being and Time* is the reductionism of the sciences of the present-at-hand in that the object reduces and loses the full sense of existentiality. Symptomatically, nature as that which "stirs and strives," as the "springhead in the dale" is lost. In the technology lecture it would seem that what is reduced and lost is the "toolshop" itself, and with it the direct expressivity which characterized the ready-to-hand of *Being and Time*. I will not deny that Heidegger provides clues himself for such an interpretation — but it seems to me that this misses much of the thrust of the Heideggerian philosophy of technology.

There is another side to the interpretation of technology which does emerge from this surface negativity. This may be seen in the transpositions which occur between *Being and Time* and the lecture. The lecture does not make anything of the distinction between the ready-to-hand and presence-at-hand, but it does elevate to the fore a strong and comprehensive concept of technology. And it seems to me that this concept is one which *combines* certain features of both the present-at-hand and the ready-to-hand in such a way that we may speak of a unique *scientific technology*. Thus, in spite of his playing down of a distinction between traditional handiwork technology and contemporary technology, Heidegger in effect recognizes the uniqueness of the latter. Indeed, the clue to the combination is not far from the surface.

If one returns to the contrast between presence-at-hand and the ready-to-hand of *Being and Time*, not only does one note the essentially positive tone which permeates the discussion of the ready-to-hand, but sees in it certain base constants which re-emerge with a different evaluation in the technology lecture. Recall several of these key features: (a) world is revealed *through* the equipmental context; (b) the equipmental context is the condition of the possibility of specific 'tools' being what they are; (c) noetically, engagement of the environment through readiness-to-hand reveals existential intentionality as *concern* (which is an index of Care in *Being and Time*), and (d) concern takes account of the context wholistically as circumspection. This *praxical* dimension is where the *essence* of Dasein is shown and effected. Now each of these elements remains constant with the later technological interpretation of the contemporary world in Heidegger, but the earlier clearly positive tone coloring these elements is transformed into the ambiguous sense of *danger* which characterizes the technological world.

Contrarily, the brief characterizations of presence-at-hand in the 'tool analysis' are often marked by partially negative characterizations. The present-at-hand originates by means of (a) deficient mode of concern (BT 103) and is characterized as a matter of entities appearing under the guise of (b) bare perceptual cognitions (BT 95), (c) just looking (BT 98), and (d) as abstract reductions interpreted as a world-stuff (BT 101). Positively, (a) presence-at-hand may be elevated into a kind of knowledge (science) which knows the world theoretically (BT 95), (b) which can be thematically ascertained (BT 106), but which appears accordingly as the ultimately reduced *functions* of the theoretically constituted world. "By reason of their Being-just-present-at-hand-and-no-more, these latter entities can have their 'properties' defined mathematically in 'functional concepts.' Ontologically, such concepts are possible only in relation to entities whose Being has the character of pure substantiality. Functional concepts are never possible except as formalized substantial concepts." (BT 122)

Now the concept of *technology* which pervades the lecture clearly *combines* elements from both sides of the earlier contrasting modes of relation. It remains the case that only through concern with the world, through what remains the praxical, is humanity effected in its essence. And it is only because it is effected in its essence that technology can be considered dangerous. "The threat to man does not come in the first instance from the potentially lethal machines and apparatus of technology. The actual threat has already afflicted man in his essence." (T 309) But what is now taken into the very way in which world is perceived are the previously negatively characterized "reductions" whereby world becomes mere standing-reserve.

I have indicated that latently the "nature" of the ready-to-hand already anticipates the notion of standing-reserve. Taking account of nature in such a way that the "wood is a forest of timber" is already to be open to a world taken as standing-reserve, but this is a necessary and not sufficient condition.

What makes it sufficient is the *addition* of thematically and systematically taking 'nature' into a calculative and *universal* view of nature as standing-reserve. But this is the metaphysics of what may be characterized as a scientific or theoretically organized technology and not that of any simple handiwork technology. Thus in some sense, the illuminating distinctions of the ready-to-hand and the present-at-hand of *Being and Time* collapse in the later work and become unified.

One result of this collapse is the elimination of any purely contemplative science. There can be no 'just looking' in what should more correctly now be called a technological science. The Greek ideal is what is lost — and if Heidegger is correct, then those who think they are remaining true to this ideal are merely naive and open to being used by technological culture. As with the non-neutrality of technology, there can now be no neutrality to science.

Ironically, a compatible way of interpreting this collapse of readiness-to-hand and presence-at-hand in the later Heidegger is to see that the science latent within presence-at-hand, in contemporary technological science, has become an *existentialized* science. That is why it can be thought of as effecting humanity in its essence. I shall not speculate concerning how this might literally be the case in contemporary genetic engineering, however tempting such an excursis might be, but it is in such examples that one might see how humanity itself becomes standing-reserve in the Heideggerian sense.

Technology, then, becomes the combined powers of what was earlier both readiness-to-hand and presence-at-hand. Humanity is effected essentially because science itself is technological in its contemporary sense and operates in the praxical dimension. But in these transpositions the earlier positive tone given to the praxical also disappears and is replaced with the characterizations of technological culture as 'dangerous', 'ambiguous', 'mysterious', and as harboring even a certain 'monstrousness'. It is from such characterizations that Heidegger's critical attitude towards technology provides material for an interpretation which sees him as dominantly pessimistic regarding humanity's future.

While Heidegger is hardly alone in this attitude, such an interpretation misses what provides not only an opening to a different hope, but the recasting of a different set of distinctions which were never fully developed in the Heideggerian corpus. I have already noted that Heidegger's hope against any totalizing closure concerning humanity lay in technics as *art*. There is a very good strategic reason for this choice.

First, art is a technics, akin thus to the concern which is exhibited in all praxical dealings with the world. It is thus already related to technology.

". . . Confrontation with [technology] must happen in a realm that is, on the one hand, akin to the essence of technology. . ." (T 317) And art is also 'theoretical' in that it does not simply take the world as that which is to be used. Its 'contemplative' attitude is thus akin to science in the earlier sense of the present-at-hand of *Being and Time*. It is interesting to note that "the botanist's plants are not the flowers of the hedgerow" (BT 100) not only contrasts with the reduced objects of a theoretically dominated presence-at-hand, but neither are they the use-sources for a sheerly praxical world. There is here a hint of a new contrast, a contrast between the now combined ready-to-hand-present-at-hand existential intentionality and the poetic being-towards-the world of Heidegger's "poetic dwelling".

Strategically, however, if artful praxis is akin to technological science in its technics and its possibility of thematic distance, its difference may also be noted. "Confrontation with [technology] must happen in a realm that is . . . on the other [hand], fundamentally different from it." (T 317) The difference lies in its *proliferation* of possibilities. Art is essentially anti-reductive in its imaginative fecundity. Its 'worlds' are effectively endless.

I am thus suggesting that in terms of Heidegger's systematic concern with praxical, now technological humanity, artful praxis is not some simple addition to the current epoch of Being, but is the strategic counter-balance to what Heidegger fears is the threat of closure. There is thus an internal need for the turn to poetics, from Heidegger's point of view, as a response to the age of technology as the current epoch of Being.

NOTE

[1] Quotations from "The Question Concerning Technology" are from *Martin Heidegger: Basic Writings*, edited by David Krell (Harper and Row, Publishers, 1977) and will be listed in the text simply as (T, pp.). Similarly, those quotations from *Being and Time* are from the John Macquarrie and Edward Robinson translation (Harper and Row, Publishers, 1962) and are listed in the text here as (BT, pp.).

CHAPTER 10

TECHNOLOGY AND THE HUMAN: HANS JONAS

There can be little doubt that Hans Jonas is among the pioneers in the philosophy of technology. That is commendable. But what is amazing in the light of the clearly obvious impact of technological culture, is that philosophy as a whole, as a discipline which prides itself in its comprehensiveness, its critical acumen and its claim to deal with the most profound human, social and value questions, should have taken until now to develop a philosophy of technology.

Yet today, philosophy *has* come to technology: All three divisions of the APA now have sections on philosophy-technology. The PSA has a section on technology. And the recent burgeoning of the medical ethics and computer fields have also run headlong into technology in the very center of their dilemmas in ethics. But what *is* the philosophy of technology? Prior to turning more directly to Jonas' philosophy of technology, I shall first attempt to give a capsule survey of what I perceive to be various approaches to the emergent field with some attention to certain key issues. In this characterization, I shall isolate three variables which may serve as lightening rods for the diverse issues which focus the current debates. The variables are these: (1) *The neutrality or non-neutrality of technology.* By this I mean that a given position may interpret technology as either neutral – it is a 'mere instrument' used by humans which has no 'life of its own'. Or it may not be neutral and have a 'life of its own' in which case at least two sub-possibilities emerge. It may be either inherently good or inherently bad with respect to a dominant direction of impact. (2) The second variable may be termed a background variable and relates to *a view of humanity's relation to Nature*. Usually this variable may be seen as some variant upon a nature–culture distinction. Here the question is one of valuing some preferred 'distance' to Nature. Is humanity's essence to be more, or less 'distant' or distinct from Nature? This variable is an exceedingly complex and unstable one in that the idea of Nature is ambiguous and can vary largely as well as can the idea of 'distance'. But it is important with respect to the question of technology in the following way: one extreme possibility of interpretation is to see humanity's essence as belonging 'closely' to Nature, in which case an interpretation of technology is to see it as the creation of an artificial realm which separates humanity from Nature. Contrarily, insofar as one might value distinctness from Nature (to value culture over nature in some versions), technology might be seen as the means of heightening this distinction and thus enhancing humanity in contrast to Nature. (3) The third variable is then a resultant variable with respect to a valuation of technology. Here both pessimism and optimism are possibilities and I think one can begin to see how these results will be arrived at. If one regards technology as neutral, and one regards 'distance' from Nature as positive, then a utopian view of technology might emerge such that it is seen as the most potent single creation of humanity to fulfill its destiny. The current problems with the environment, etc., would then be interpreted as either temporary problems due to the primitive state of current development, or as mere bad uses of technology, blamable on some bad social or economic organization. Contrarily, if one regards 'nearness' to Nature as positive and technology as non-neutral (in its negative sense), then one will arrive at a pessimistic view in which technology poses the very most profound threat to the very essence of being human.

Admittedly, these variables are sketched in the starkest terms for the sake of clarity, but I believe that they will help lend insight into the current status of the philosophy of technology. I shall now turn briefly to the current, actual situation. I note that to date there seem to be approximately five identifiable groups which are working on what is becoming the philosophy of technology:

(A) The first and probably most sustained group is roughly Marxian in origin. Here the emphasis has been upon, first, an examination of the concrete modes of production in which the question of technology must be raised, and then upon the social organization and uses to which these are put. This group is *not* monolithic with respect to the variables sketched as there are utopian marxists at one extreme, while others stress the problem of alienation and are sometimes quite pessimistic (Marcuse, for example). But note three things about this group: (i) it does look at technology in terms of its concreteness, its actual use and development within a social context. (ii) It raises the question of alienation, but alienation usually takes the special meaning of a distance between the producer (creator) and his products (goods), and not a distance between humanity and Nature. In fact, most of the marxian emphasis is one which 'distances' humanity from (raw) Nature. (iii) This group vacillates with respect to the neutrality or non-neutrality of technology, but it inclines towards minimal non-neutrality in that concrete modes of production are seen to have effects which far exceed the specific intentions or purposes to which they are put.

(B) A second grouping is equally diverse and I shall call them the theo-existentialists. It combines both religious roots with more modern existentialist themes. With respect to the variables this group usually sees technology as non-neutral. But it rarely sees non-neutrality positively although there are some exceptions (Mounier, for example). It also deals with the concept of alienation, but in this case alienation is not so much a 'distance' between humans and the products of their work as between humans and their essence, a variant upon the notion of Nature. The view of technology here is usually negative and is seen as both something which intrudes between humanity and its essence and as something like a Frankenstein which, once created, outruns and threatens its creator.

(C) A third group, often less self-conscious about a philosophy of technology, may be identified amidst the analytic philosophies. The most interesting group here is one which works on artificial intelligence. Coming from questions in the philosophy of mind and out of a contemporary version of a dualistic materialism, it usually sees technology as neutral, but insofar as the brain *is* conceived to be a kind of machine, humans are thought to be already 'close' to a mechanistic Nature. Thus the outcome is usually optimistic and utopian. I shall call these the Mechanists.

(D) A fourth group is very diverse and may be termed the Applied Philosophy group. Here the approach is a traditional one in that technology is rarely viewed for itself, but interrogated with respect to its effects. The ethical implications, questions of distributive justice, and the clarification of problems within the context of decision or systems theory are the most important. Typically, utilitarians may be found here. Many philosophers of science may also be grouped here. With respect to the variables: technology is usually thought neutral, the mere result of human inventions and purposes; there is usually lacking any alienation theory on either marxian or theo-existentialist grounds, and thus the judgment upon technology is rarely either strongly negative or positive. The main question arrives around the uses and the epistemological, ethical and social problems of the effects of technology.

(E) The fifth group I shall call the Phenomeno-Ontologists and they often overlap with several of the above groups. This group probably is most influenced by another pioneer in the philosophy of technology, Martin Heidegger. With respect to the variables this group holds that technology is non-neutral, but this non-neutrality while often seen to be negative, relates to the Nature question in an odd way since technique is seen as a form of life or mode of being and hence is one possible essential way of being. While most heideggerians are thought to be negative with respect to technology, this is not a necessary implication of the position. But like the marxians and the theo-existentialists, the phenomeno-ontologists do directly address the theme of technology in a concrete sense. It is interesting to note that the organized, self-conscious philosophy of technology groups are currently dominated by groups (A), (B) and (E) and thus the tone is usually critical or negative regarding technological civilization. It is also the case that each of these groups have roots in European Romanticism.

Having now a set of variables to watch and having characterized some extant groupings of philosophers of technology, I shall now turn to Jonas. Long before it became fashionable, he looked at (i) the total impact of technology upon civilization; (ii) he specifically inquired into technology and the biological realm, including an examination of human death and the ethics of life and death; (iii) he has raised the question of ethics (in a non-utilitarian context) with respect to technology's personal, social and civilizational impact; (iv) and he has questioned over how or whether technological being can be a new mode of human being.

Thus it would seem that in certain ways Jonas' work is considerably broader than any of the specific groupings I have sketched. But he obviously belongs much closer to the marxian, theo-existentialist and phenomeno-ontological European groups which deal with technology concretely than to the others. I, myself, will happily admit to being an avid reader of Jonas and to having used his books in my own primitive attempts at formulating aspects of the philosophy of technology with respect to courses on technology and the environment.

But the purpose here is not merely to extoll — it is also to be critical. This is because Jonas simultaneously seems so plausible in his attitude towards the impact of technology, and yet also, to my mind, also is open to serious questions. My ultimate theme will be precisely one which lies central to Jonas' thought: the relation of the essence of man to technology with emphasis upon its ethical impact.

My first task, then is to locate Jonas with respect to the variables previously sketched. The location, I will admit, cannot avoid some over-simplification, but I think it will also be clear with respect to general conclusions for readers of Jonas.

(1) With respect to the question of the neutrality or non-neutrality of technology, I think it fair to say that Jonas falls among those who regard it as non-neutral. In his recent, provocative article, "The Ethics of an Endangered Future", Jonas argues that modern technology is essentially different from classical technology prior to the scientific revolution and that the

Baconian ideal which links knowledge to the *control* of nature was neither able to foresee nor able to control the outcome of what he calls a "third power" within technology:

> Bacon did not anticipate this profound paradox of the power derived from knowledge: that it leads indeed to some sort of domination over nature (that is, her intensified utilization), but at the same time to the most complete subjugation under itself. The power has become self-acting, while its promise has turned into threat, its prospect of salvation into apocalypse.[1]

There can be little doubt about the non-neutrality implied here. Technology, originally created by humans, once brought into being, fostered and nurtured, takes on increasing powers *vis-à-vis* nature, until it is feared that a peculiar kind of dominance over Nature will in fact be attained. Once again Golem *cum* Frankenstein emerges to threaten the human who has the hubris to have dared to be God.

(2) The overwhelming conclusion which arises out of this philosophy of technology is one which sees in its contemporary form, a threat to the very essence of humanity. The conclusion is pessimistic. The same article elucidates an ethic built upon a 'heuristic of fear' designed to awaken humanity to the threat posed by technology. About that I shall have more to say, but at the moment the question is one of placing Jonas amidst the variables.

(3) Given the way the variables relate to one another, one may surmise that if technology is perceived as non-neutral and if the conclusion is a pessimistic one, that the rationale will likely be one which values some version of 'closeness' to Nature in some sense that humanity is seen as being increasingly alienated from its source or essence. That, too, may be read rather clearly in Jonas' position. Again in the same article,

> There is no need, however to debate the relative claims of nature and man when it comes to the survival of either, for in this ultimate issue their causes converge from the human angle itself... Since, in fact, the two cannot be separated without making a caricature of the human likeness... Such narrowness in the name of man, which is ready to sacrifice the rest of nature to his purported needs, can only result in the dehumanization of man, the atrophy of his essence even in the lucky case of biological preservation.[2]

Thus the overall, first impression, particularly with respect to the three variables, leads one to locate Jonas primarily and dominantly within the group I have called the theo-existentialists. (Note that I am not here dealing with the possible truth or falsity of the position — I am merely characterizing it. But equally, I do hold that this position is open to debate.)

While ultimately and grossly, I believe this location of Jonas with respect to the key variables is fair and correct as an interpretation, it also leaves out some more subtle specific issues which need to be looked at. But even in the gross form, this location can point up two themes which must be addressed if Jonas is to be judged adequately: (a) First, the most central issue is one which revolves around what form the Nature argument takes. In Jonas' language it revolves around the notion of the "essence of man". Insofar as technology can either alienate, threaten or distort the essence of man, it can be a genuine threat (over and above the bottom line which is, of course, biological extinction). About the essence of man and its relation to the development of technology, Jonas has a good deal to say. (b) But inextricably bound to this question is another of high import as well — what is the "essence' of technology? Or, what is the phenomenon of technology and how is it bound to humanity? Here, it seems to me, Jonas is less explicit except in a highly suggestive way. (c) It is only when both of these questions are clear that an ethics may emerge. It is in the imbalance that I shall locate what I believe to be serious problems with Jonas' approach.

Even more specifically, there are two vectors which I should like to examine in Jonas' work: (a) the relation of the "essence of man" to Nature with its purported ethical implication and (b) the distortion caused by the rise of technology, again with an eye to the ethical implication.

First, the Nature—essence of man relation: As indicated previously, one problem which always emerges with the Nature—essence of man relation is one which revolves around the ambiguity of what shall count as Nature. An examination of this concept in Jonas indicates what I shall describe as a continuum of levels ranging from 'bare' or minimal Nature to 'total' Nature. The concept is important because Jonas' ethic holds that man is in some way responsible for Nature. But first, take note of the continuum of levels involved in the concept: (i) The one extreme is Nature in its widest sense as a *bare, but necessary condition of human life*. However, such a Nature could well also be a minimal Nature, i.e., just that which could be necessary for the sustenance of life. Jonas imagines such a nature in "(science fiction style) . . . a human life wòrthy of its name . . . [is] imaginable in a depleted nature mostly replaced by art" and concludes that such a minimal condition would "still hold . . . the plenitude of life, evolved in aeons of creative toil and now delivered to our hands, [it] has a claim to our care in its own right."[3]

Now while such a vision, purposely painted in bleak tones, is hardly appealing, it does carry the obvious bite that if reduced further the very condition of human life would disappear. Thus the strongest sense of self and societal interest in self-preservation appears at this level — to care for such a

Nature is *ipso facto* to care for ourselves and to neglect it is to commit self and societal suicide.

Somewhat higher up on the scale lie various degrees of more enriched notions of nature which include the fuller realms of various animal life and other organic constituents which enrich Nature to which Jonas argues we also owe care. At this level more is involved than more self-interest, but something like an enrichment of human being. "If the prerogative of man were still insisted upon as absolute, it would now have to include a duty toward nature as both a condition of his own survival and an integral compliment of his unstunted being."[4] Here the direction is clear as is the implication: the more man respects and cares for Nature, the less 'stunted' his being. But, note the grounds for an ethical argument also shift. Nature is here more than a condition for being, it is also what I shall call 'aesthetic' in that it enriches human being. But such a shift also begins to lack the 'bite' of the lower lever of necessity. It is an appeal to relative richness or variety.

The top of the scale may be predictable from the previous two levels: the full *dignity* of humanity comes about only when humans see that, "we owe allegiance to the kindred *total* of [nature's] creations."[5] There is nothing lacking in commendability in this notion – its source in the Spinozistic, Franciscan and other 'Chain of Being' roots of our religious traditions are obvious enough. But what I wish to point to is that the higher on the 'aesthetic' scale and the farther from the level of 'necessity' the less obvious the bite of the ethical demand. Thus while each of us might well agree that we *ought* to do something to protect the endangered snow leopard, it becomes very unclear that its extinction would do much to specifically lessen or clearly endanger the *conditions* of human life since the effect is likely to be little detectable. But, contrarily, for the air to reach a saturation point in pollution beyond which health and even life is endangered, can more easily come into the region in which Jonas' notion of necessary care for Nature may have bite. In short, I am pointing up what seems to me to be a shift in level concerning our responsibility to Nature insofar as it is a condition of human life. I shall address this shift to Jonas' notion of a "heuristic of fear" in a moment.

If, now, we combine the notion of a shift in levels with the introduction of a shift in the essence of man, the following problematic emerges. First, there is the possibility of a conflation of the notion of duty which man owes to Nature between the bottom line and the top line. To owe duty to necessary Nature is to have the bite of self and societal self-preservation; but to owe the duty to enriched Nature is to have the appeal of aesthetic richness. The bottom claim bites; the top appeals and is further more debatable in that what one takes from the richness may vary considerably. (For example, I might want to argue that the *total* elimination of certain viruses, if made possible technologically, would enhance the essence of man. This would be an argument *against* the duty to the totality of Nature's creations.) This possible conflation of levels must be kept in mind when we turn to the second notion which introduces technology.

The second key notion I wish to address is the idea of a *'new* ethics' which arise from what is noted by Jonas as a distortion within the "essence of man" caused by the rise of technology. Jonas, in effect, argues that there has been a *change* in the contemporary situation of life and by implication raises the possibility that the essence of man may have suffered a change due to modern technology. But at the same time this change is seen *negatively* and may be interpreted as a *distortion* of the full essence of man.

In an earlier essay, "Technology and Responsibility: Reflections on the New Tasks of Ethics", Jonas argues that until contemporary times it was possible to hold that "the human condition, determined by the nature of man and the nature of things, was given once for all,"[6] but that now:

> These premises no longer hold. . . More specifically, it will be my contention that with certain developments of our powers the nature of human action has changed, and since eithics is concerned with action, it should follow that the changed nature of human action calls for a change in ethics as well.[7]

This change brings about a more radical result:

> [a] qualitatively novel nature of certain of our actions has opened up a whole new dimension of ethical relevance for which there is no precedent in the standards and canons of traditional ethics. The novel powers I have in mind are, of course, those of modern technology.[8]

I have earlier referred to his claim which holds that these powers have now been escalated to the third power such that even the Baconian sense of knowledge as power has no power over the released and dominantly negative technological development of the times.

The relevance of this change for the "essence of man," this novel and negative condition, is one which Jonas indicates *distorts* the essence of man:

> [Technology's] cumulative creation, the expanding artificial environment, continuously reinforces the particular powers in man that created it, by compelling their unceasing inventive employment in its management and further advance, and by rewarding them with additional success – which only adds to the relentless claim. This positive feedback of functional necessity and regard – assures the growing ascendance of one side of man's nature over all the others, and inevitably at their expense. . . The expansion of his power

is accompanied by a contraction of his self-conception and being. In the image he entertains of himself – the potent self-formula which determines his actual being as much as it reflects it – man now is evermore the maker of what he has made and the doer of what he can do, and most of all the preparer of what he will be able to do next.[9]

The specific distortion, then, caused by technology is to have enhanced the powers of man which must genuinely be said to be part of his 'essence' at the expense of his self-conception. But once unleashed, this distortion now has gotten out of control and all previous and normal ethical formulae will be impotent before the distortion.

This sounds amazingly like a contemporary 'fall of Adam' in which case the role of serpent is played by the seductiveness of the machine. I suggest this purposely because the next step is to look at the proposed remedy which is a *new* ethics motivated by a 'heuristics of fear'. The situation today, like that after the Fall, is one in which once begun, sin (read technology) has gotten out of hand. And what is needed is a strong remedy to return to the original state of Nature in both its strong and weak sense. Jonas argues, "We need today an imaginative-anticipatory heuristics of fear to lead us to the discovery of the duties, even the principles, with which to meet the challenge of coming events."[10] I shall not go deeply into the argument by which this 'heuristics' is justified, other than to identify it as a dialectical one. It is, in fact, what we used to know in graduate school as the 'aesthetic' argument which was a variant upon the question of could there be evil if God was truly good? The aesthetic argument went: we would not know good were there not its opposite, evil, and thus to be both good and knowing there must be evil. (The rejoinder was, of course, if this is true, why have so much evil to make the point?)

Jonas' version is indicated as follows:

Just as we should not know about the sanctity of life without knowing about killing and without the commandment "Thou shalt not kill" having brought this sanctity into focus; and just as we should not know the value of truth without being aware of lies, nor of freedom without the lack of it, and so forth – just so in our search after an ethics of responsibility for distant contingencies can an anticipated *distortion* of man help us to detect that in the conception of man which is to be preserved from it... We need the *threat* to the image of man ... to insure ourselves of his true image by the very recoil from these threats.[11]

I contend that the heuristics of fear which arises from this dialectic – that we must face *total* threat before we can feel responsibility for *total* Nature and that fear is the stronger motivation for this responsibility, is far from new. It is, in fact a revival of one of the oldest ethical traditions in the West and carries with it in new guise the old notions of a Fall of man and his recuperation through some form of salvation, either by Law and fear or by Faith and Grace.

I would go further, I think the heuristics of fear with its clearly more strongly weighted emphasis upon "perception of the *malum* as infinitely easier"[12] its "prophesy of doom is to be given greater heed than the prophecy of bliss."[13] and its clearly apocalyptic context is one which returns us in new clothing right back to Fall, God and the Devil. It is an ethics which is not only based upon an ethics of fear, but is a fearful ethics. It is an ethics which in response to what Jonas perceives as the heightened ambiguity of the contemporary situation, the heightened complexity of the human situation, and the heightened uncertainty of more and more long term projections, reverts to what I shall call a strategy of conservativism which essentially advises: if we cannot make long-term projections in the light of uncertainty, it is doubtful that we ought to undertake the actions beyond a certain magnitude at all.

Jonas states this as: "Never must the existence or the essence of man as a whole be made a stake in the hazards of action. It follows directly that bare possibilities of the designated order are to be regarded as unacceptable risks which no opposing possibilities can render more acceptable."[14] But here we return to the previous confusion of conflated levels. What decision could possibly risk the whole essence of man? If we mean the bare existence of man in minimal nature, then such technologies which might threaten the entire biosphere (nuclear warfare and destruction of oxygen producants in the ocean) might well fall under the question of the strategy of conservation. But if we mean the unstunted, enriched essence of man with respect to the toality of Nature, then we mean any technology at all! This is because any technology raises a question with respect to man's essence. It is here that the near absolutism of this new, old ethic begins to come clear. Ultimately, it is an apocalyptic ethic which as with all apocalyses poses a clear either/or.

What I have argued to this point is not meant to undercut everything Jonas has done in the philosophy of technology. It is meant to isolate what I believe is a *serious conceptual* problem relating to the ethics which is to be the import of that attitude. And even in my lack of sympathy towards an ethics which emphasizes a heuristics of fear, I find a suggestion which merits a much longer consideration. *If* technology can change the essence of man — even if this change is interpreted as a distortion by Jonas — there is room for development. I cannot pursue this here, although I should like to suggest that it lies more centrally in the trajectory taken by the phenomeno-ontologists, but

I would like to pose it in terms of two questions:

1) Could it be that technology itself *is* an expression of the essence of humanity, not merely in a distorted sense, but in all the ambiguity found in man? 2) And, if so, what is the phenomenon of technology such that it so clearly *amplifies* the very possibilities of that humanity such that man may become a threat to himself?

NOTES

[1] Hans Jonas, 'Responsibility Today: The Ethics of an Endangered Future', *Social Research* **43** (1) spring, 1976, p. 84.
[2] *Ibid.*, pp. 77–8.
[3] *Ibid.*, p. 77.
[4] *Ibid.*, p. 79.
[5] *Ibid.*, p. 78.
[6] Hans Jonas, *Philosophical Essays* (New York: Prentice Hall, Inc., 1974), p. 3.
[7] *Ibid.*, p. 3.
[8] *Ibid.*, pp. 3–4.
[9] *Ibid.*, p. 11.
[10] Jonas, 'Endangered Future', p. 87.
[11] *Ibid.*, p. 87.

CHAPTER 11

THE SECULAR CITY AND THE EXISTENTIALISTS

Theology in a popular sense has begun to come to honest grips with the technological and secular culture of our time. This is evidenced in the wide reading accorded to Harvey Cox's *The Secular City*. A frequently repeated reaction to *The Secular City* is that it leaves one with a sense of having been freed — but without the reader being able to say just what it is that he has been freed from.

I wish to specify what I believe to be one dimension to this sense of freedom which seems to result from reading *The Secular City*. Cox, without ever making it totally explicit, has performed an iconoclastic exposing of one of the social myths of our time.[1] This myth may be called in general an antitechnological myth.

In my own reading of *The Secular City* I find myself in basic sympathy with Cox's position and the argument which I wish to enter with him is a friendly one, an argument which hopefully can add to the discussion of secularization. What bothers me is Cox's polemic against the 'existentialists.' Cox never makes clear just who these 'existentialists' are in his attacks. This is all the more confusing since Cox draws with obvious approval from Albert Camus, the early darling of American readers of existentialist literature, and from Merleau-Ponty, now one of the most studied philosophers for students of existentialism. Cox even seems to approve of one of Simone de Beauvoir's insights and exempts Sartre from some of his charges.[2] This attack distresses me as a student of existential philosophy, not only in its too general impetus which confuses, but because, as I hope to show, it is just plain wrong as a general attack.

Most of the attack is mere polemic through general swipes at 'existentialism' as a 'philosophy of meaninglessness';[3] as 'irrelevant' to the new epoch;[4] and as 'immature.'[5] But these often repeated and ill-founded terms are merely name calling. The most specific attack comes in almost the form of the antitechnological myth which I contend Cox is exposing in *The Secular City*. In Cox's charge it almost appears that the existentialist philosophers are the group who are most responsible for its continued social power:

Existentialism appeared just as the Western metaphysical tradition, whose social base was dismantled by revolution and technology, reached its end phase. It is the last child of a cultural epoch, born in its mother's senility. This is why existentialist writers seem so arcadian and antiurban. They represent an epoch marked for extinction. Consequently their thinking tends to be *antitechnological, individualistic, romantic*, and deeply suspicious of cities and science [italics mine].[6]

It may be the case that *certain* existentialist writers do tend in that direction or do have elements in their writing which could be so construed. Jaspers, for example, does have an antitechnological bias, yet he is not mentioned in *The Secular City*. It is also clear that Heidegger is a bit romantic about some of the presocratics, and Cox attacks him for this backward looking. (It is interesting to note that some "secular" readers of Cox are perplexed by Cox's own "backward look" to the Bible for insight into the current situation. Nor is a backward look necessarily merely that, for one should recall that the Reformers looked back to former times, as did the leaders of the Renaissance. Backward looks often bring about something new rather than a mere revival of the old.) Further, some scholars of Heidegger would remind Cox that it was Heidegger, after all, who helped much of the philosophical world become more *history conscious.*

Nor does Cox seem to consider the fact that existentialism as a philosophical movement has not stood still. If Sartre's *Being and Nothingness* seems overly "individualistic," it must be remembered that this book is now balanced by his *Critique de la raison dialectique.*

Although it would be possible to continue in line with the above, it appears more fruitful to me to add to the discussion of secularism while showing at the same time that not all philosophies of existence come under Cox's charge. In fact, the two thinkers I wish to comment upon anticipate many of the sound points made by *The Secular City*. Neither of the two can be considered antitechnological, romantic, or individualistic.

In 1932 there appeared in France a new review, *Esprit*, founded by Emmanuel Mounier, who had left his teaching post to undertake the financially risky task of heading a new intellectual movement dedicated to the discussion and renewal of man's self understanding in a self-consciously modern sense. Around Mounier and his journal a number of Christian intellectuals gathered, the best known of whom was Nicolas Berdyaev, who is usually included in anthologies of existentialist writers. Another was Paul Ricoeur, who was to become the Sorbonne's chair of philosophy professor following Merleau-Ponty's death. It is with Mounier and Ricoeur that I wish to deal here.[7]

Now to set up the dialogue between Cox and his 'existentialist' anticipators in such a way as to belie his charges and at the same time add to the discussion which exposes the antitechnological myth. Cox's charges are that existentialism is: (a) antitechnological; (b) romantic; and (c) individualistic. Both Mounier and Ricoeur would agree with Cox's negative valuation of this reaction to the secular.

Emmanuel Mounier (1905–50) makes it clear that this reaction has the dimensions of a complex and well-functioning social myth. He calls this the 'anti-machine myth.'[8] On one level this distrust and fear of the machine, now being reasserted in certain quarters by attitudes towards computers, appears to arise over a fear of the social role of the machine. Mounier recognizes and does not play down the sometimes negative social consequences of the machine, particularly in the use of new instruments of war, but also through the application of new machines which change the productive process. Such changes with socially negative results serve to re-enforce the continuance of the anti-machine myth.

But to this level of the analysis Mounier appreciates and applies the Marxian insight:

As a matter of fact the anti-machinist myth started as a bourgeois myth: the organized working class, after a few upheavals at the beginning of the industrial revolution, remained solidly aloof. It is understandable. The worker knows his machine, it does not secrete for him the horror of the unknown or any loss of prestige, and under the influence of Socialist teaching, he soon learnt to attribute is misdeeds, not to the nature of the machine, but to the use that was made of it.[9]

The social system which uses machines to exploit may be more complex than the social system which employs human slavery for the same reasons, but it is certainly no worse. One might add that the society which refuses to adjust and redistribute its wealth with the advent of new machines can hardly blame the machine for its recalcitrance.

Secondly, part of the negative reaction, particularly to new machines and the vast proliferation of invention in recent decades, Mounier contends, is due to our inability to attain perspective.

As far as the machine is concerned, we are collectively at a moment of disorganization. It is a very short term view to believe that our actions will get more and more uncoordinated and stupid until we reach catastrophe.... Today the machine is not adjusted to the rhythm of man. There is no law that the two should be in eternal disharmony.... We simply forget that it has taken several hundred thousand years to establish and consolidate this rhythm: man as a maker of machines has had two centuries.[10]

But the antitechnological attitude has deeper roots in human history than merely negative reactions to immediate problems and the lack of perspective. The antitechnological attitude does not stand alone, but is, at base, connected to the affirmations of a type of romanticism. Mounier recognizes this and indicates that it is here that the social myth takes on its nearly religious dimensions.[11] Recent versions of this romanticism draw their strength from myths deeply rooted in man's collective symbolism, a symbolism which was once that of Nature as our eternal mother.[12]

The romantic understanding of Nature is one in which man is not only intimately linked with Nature, but he is seen to have a stable and unchanging nature which provides a more comfortable understanding than a more radical philosophy of human freedom. With romanticism, "this nature ... was once humanized in myth: gods and goddesses converted this strange absolute into a mirror eternally reflecting back to man his own reassuring image."[13] It is out of this fertile myth-ground that modern romantics forge their opposition of the 'natural' to the 'artificial.' It is from this instinct, for myth, Mounier claims, is closer to instinct than to reason, that the nostalgia for a return to Nature or a simpler past arises.

Mounier sees in Judaeo-Christian thought a clear opposition to this myth and, as Cox later does, links Christian intentionality to that of modern science. "The resistance of folklore to Christian monotheism shows how greatly man needed this fraternity of the universe in order to struggle against the vertigo of its dimensions and its silences. This protective system disappeared under the joint attack of monotheism and modern science."[14]

If antitechnological romanticism is seen negatively by Mounier, then a more positive attitude toward technology is affirmed by him. Science and technology, even in a most frightening result, the invention of the H-bomb, spell for Mounier a new recognition of the freedom and responsibility man has for his destiny.

> We have acquired a unique power, the reverse of all others, the power to explode this planet, with all humanity and its power to create power. It is a dramatic moment. We could not truly say until now that humanity was master of its future, since it was still condemned to have a future, whereas each individual can, if he so wishes, put a pistol to his head. But now humanity as such has to choose and, as far as we can see, it will need a heroic effect not to choose the easy way of suicide. One might say that humanity has now become adult.[15]

Technology, far from being anti-human, is thoroughly human. Against romantics who decry the 'artificial' Mounier replies, "Are the books of St. Exupery, who wrote about aviation, less human than those of M. Henri Pourrat, who wrote on ploughing?"[16] On a higher level technology as a human venture challenges man to undergo an adventure:

> The machine was the first step towards adventure, but we tried to tame it by making it a promise of happiness. Happiness was not the result, and we naively reproached it for

letting us down, as if happiness were to be provided by the machine or anything else on earth. But this disappointment is still a refuge and a way of keeping one's gaze upon the past. Modern man is able to admit his real anguish more openly; whereas the machine offers him the mastery of his kingdom, albeit in struggle and at great risk, man is afraid that it will snatch his empire from him, that apparent sovereignty that existed only in the long immobility of things.[17]

These 'small fears' of the twentieth century, as Mounier called them, need to be seen not in the light of a fear of the rape of nature, but in the light of the adulthood of man. If man is going to come to grips with his inventions he must continue the struggle and gain an understanding of the logic of the machine. Mounier reminds us of an earlier day in the history of technology. Once engineers and designers decorated their engines with baroque trimmings or tried to get machines to imitate some natural thing — this ignores the form of the machine as such. "But it so happens that the machines that are the most clumsy and inefficient and propertyless are precisely those that bear the closest resemblance to natural objects."[18]

Mounier's own understanding of the logic of the machine is not far developed. However, in the few suggestions he makes one finds a positive spirit. He recognizes, for example, that it is possible for a technocracy to emerge. But while the future remains ambiguous there is reason for hope even if such a form of society were to happen:

A technocratic imperialism, however, is governed by the laws of efficiency: the least kindly technocrat would introduce socialism if he thought it would improve the organization and the least humanitarian would hesitate to start a war that did not pay. The machine therefore, has some form of reason to counter its unreason.[19]

More positive is the suggestion, not original with Mounier and now much discussed with the rise of computing machinery, that we ought to understand the machine not merely as an extension of our physical limbs, but on the analogy of our language:

The machine as implement is not a simple material extension of our members. It is of another order, an annex to our language, an auxiliary language to mathematics, a means of penetrating, dissecting and revealing the secret of things, their implicit intentions, their unemployed capacities. Language, when it gives precision to an idea, can be accused of hardening and immobilising it. Should we therefore be inarticulate?[20]

To this point I have stressed Mounier's contribution to a discussion of the growth of technological culture, a discussion which is in sympathy and which predates Cox's *Secular City*. If one were to read further, or even see between the lines of the small selections quoted here he would not have to go far to find a reiteration of certain existentialist themes. Mounier recognizes the *vertigo* man faces due to the fact that man's fundamental concepts are always being called into question.[21] But vertigo must merely be accepted as part of the task of humanity's adulthood. Mounier certainly sees the *ambiguity* of the human situation in a technological world, a world in which there is no 'unseen hand' guiding its destiny as such. But Cox, too, rejects the simple providentialism of the past. Mounier sees technology as an important part of the human situation which raises in dramatic fashion the questions of human *freedom* and *responsibility* in a way which makes man aware of his *choices* and his *future* as never before.

None of this 'existentialism' brings Mounier to an antitechnological or romantic view of our epoch. Rather, it brings the challenge which is proper to the adulthood of man.

With Mounier I have primarily addressed the issues of romanticism and an antitechnological attitude. The third charge, individualism, I have left for answer to Paul Ricoeur.[22] Again, however, the discussion may be resumed within the context of the topic of technological culture.[23] Ricoeur continues to see technology, though in a broader sense, as a primary factor in man's emergence and distinction from nature.

If man stands out in such sharp contrast to nature and the unending repetition of animal habits, if man has a history, it is primarily because he *works* and because he works with *tools*. With tools and the work produced from them we touch upon the striking phenomenon that the tool and its products are preserved and accumulate. Thus we have here a phenomenon which is truly irreversible.[24]

One should note here that Ricoeur's understanding of tools is a rather broad one, for not only do material tools count as human tools, but also all the instruments by which man may accumulate results:

The technical world of material tools and their extension into machines is not the whole of man's *instrumental world*. Knowledge is also a tool or instrument. Everything man has learned and all that he knows – everything he can think, say, feel, and do – all that is 'acquired.' Knowledge becomes stratified, deposits of knowledge accumulate like tools and the works which result from them. Concretely, it is writing and especially printing which have permitted knowledge to accumulate and leave traces.[25]

Here 'technology' is seen as a rather broad dimension of human activity and an obviously important dimension.

Man's historical existence, in which the accumulation brought about by work with tools, implies for Ricoeur a separation of man from nature which is positively valued. Speaking of a Christian meaning of history in the present Ricoeur holds:

It would seem that the value found at this level is the conviction that man fulfills his *destiny* through this technical, intellectual, cultural, and spiritual experience; yes, that man fulfills his role as a creature when he breaks away from the repetition of nature and makes his history, integrating nature itself into his history and pursuing the vast enterprise of the humanization of nature.[26]

Thus for Ricoeur the primary problem of the time is not to be solved by romantic or antitechnological attitudes. On the contrary, he specifically denounces those who would destroy the autonomy of scientific investigation as such: "The spirit of truth will not criticize the dehumanization of man by scientific objectivity...."[27]

Science and technology and their development must be protected in their relatively autonomous level of truth. Where they become a wider and more specific human problem is where they cross the ethical and social existence of man. Man's ethical existence cuts across all the specific activities in which he engages. In a discussion of "The Socius and the Neighbor" Ricoeur argues:

The vice of the social existence of modern man does not lie in being against nature; what is lacking is not naturalness, but charity. Consequently, criticism goes completely astray when it attacks the gigantism of industrial, social, or political machinery, as if there were a 'human scale' inscribed within man's nature.... We are in need of a critique other than this idea of Greek 'measure' which opposes the great planning researches of modern social life. Man's technical, social, and political experience cannot be limited in its extension....[28]

Nor is charity understood by Ricoeur to be a merely individualistic notion. It may well be that it is out of intimate relations that one may first understand the depth of love, but that does not exhaust one's capacity or responsibility.

It is with the same emotion that I love my children *and* [italics mine] take an active interest in juvenile delinquency. The first love is intimate and subjective, albeit exclusive; the second is abstract but has a wider scope. I am not discharged of all responsibility to other children by simply loving my own. I cannot escape others, for although I do not love them as my own or as individuals, still I love them in a certain collective and statistical manner.[29]

The individual and the socius are always bound together. Even the intimacy of the family, concrete as it may be, cannot exist apart from its context: "The family home has no intimacy unless shielded by legality, a state of peacefulness based on law and force.... The abstract is what protects the concrete, the social establishes the private."[30]

And because this is the case the ethical and social demands placed upon an ethics of charity by the neighbor cannot be restricted to one-to-one relationships:

> The object of charity quite often appears only on the form of a collective misfortune: wages, colonial exploitation, racial discrimination. Then my neighbor is concrete in the plural and abstract in the singular: charity reaches its object only by embracing a certain suffering body. . . . It is not necessary to enclose oneself within the letter of the parable of the good Samaritan, nor to impose upon it a personalist anarchism.[31]

If the lines between the socius and the individual intertwine in ways which forbid 'personalist anarchism' and by the same token a dogmatic collectivism (Ricoeur is sympathetically critical of Marxism for this reason), then it still remains to see just where one is to find the crossing of these lines. Ricoeur's answer is clearly drawn from the phenomenological-existentialist tradition in the concept of the world of perception or life-world. In discussing the impact of certain modern problems he indicates:

> The military, industrial, and economic problems which are consequent upon the discovery of atomic energy are not raised in terms of the truth or falsehood of the atomic theory, but rather in terms of our existence; they are raised in the world such as it appears. They do not come up in the universe such as the physicist represents it to himself, but in the world of perception in which we are *born*, in which we live and die. It is within the world of perception that our instruments, our machines have an ethical significance and bring into play our responsibilities.[32]

At this point the limits of a contrived discussion between Cox and two existentialist sympathizers are reached. With Ricoeur the charge of excessive individualism should be laid to rest. What remains is a short final clarification of the issue surrounding the antitechnological myth Cox is exorcising.

I have not denied the existence of a set of attitudes which may be characterized as antitechnological, individualistic, and romantic. I have denied that the existentialists in general should carry the blame for supporting these attitudes. If these attitudes constitute a functioning social myth, I suspect they are more general and less to be found in a specific group than appears to be the case in Cox's charge.

I further suspect that Cox's 'existentialists' may be more recognizably some American readers of existentialism who select from what they read that which fits an already preconceived pattern which approximates the attitude Cox condemns. They are like the freshmen about whom a graduate assistant remarked, "They like the existentialists, but they see only the freedom without seeing its heavy responsibility; they want to be rugged individualists, but they have not grasped *Angst* and commitment."

There has been in recent time a rather wide-ranging literature which tends to support the antitechnological (and now we must add, anti-organizational) myth. It runs the gamut from the 'organization man' literature to the much discussed problems of 'alienation' in departments of sociology.

Ironically I found a large number of students in a most technological setting, the Massachusetts Institute of Technology who rather well illustrate what a believer in the myth might hold (provided one merely substitute 'anti-technology' with 'anti-organization'). For in spite of the fact that MIT's laboratories and projects are primarily supported by funds and the desires of the 'establishment,' i.e., the government and large organized industry, the cry from many was that the ills of our time lie in a loss of individualism and the growth of federal power. There was in addition a certain nostalgia for a (probably mythical) day when scientists were able to make discoveries all by themselves in their very own laboratories. All of this was expressed in the very place in which it should have been obvious that the glory of modern science and technology is not possible apart from gigantic funding and organization.

If Cox's point is that it is time that we face, accept and deal with technological and organizational culture as a now given fact of our existence and do so as adults, then I must agree with him. The task remains the same in essence as it was when man first began to emerge from nature: How does one use what one finds and what one creates to fulfill humanity? If that question must now be addressed primarily to the organization and the social giantism we have created, only its object has changed. What is needed is an awareness of this with the search for and development of methods appropriate to the object. Finally, if the exorcising of an antitechnological myth is needed to accomplish this, then more power to both Cox and his existentialist anticipators.

NOTES

[1] By myth I intend, not a mere falsehood, but a set of directive ideas which serve to set rather basic attitudes in our understanding of our existence.

[2] Harvey Cox, *The Secular City* (New York: The MacMillan Co., 1965). Reference to de Beauvoir, p. 195; to Sartre, p. 252.

[3] *Ibid.*, p. 68.

[4] *Ibid.*, p. 80.

[5] *Ibid.*, p. 253.

[6] *Ibid.*, p. 252.

[7] I first became interested in this group in 1959 while doing a thesis on Berdyaev. Later the interest continued with a dissertation on Ricoeur. The *Esprit* circle, it is true, never explicitly endorsed existentialism as a title (neither did any major philosopher usually associated with existentialism except Sartre. And even Sartre now denies the title.) Mounier's philosophy began as "Personalism" in which, however, the person was

contrasted with the mere individual and always seen as a part of a community. Later, in his *Introduction to Existentialism* and towards the end of his life, he moved to a closer affiliation with existentialism as such. Ricoeur, one of France's leading scholars of phenomenology, has in his major work, *Philosophie de la volonte*, elaborated a philosophic anthropology of man as a broken unity which has obvious connections with existentialist themes.

[8] Emmanuel Mounier, *Be Not Afraid*, trans. Cynthia Rowland (London: Rockliff, 1951), p. 30. This book is a collection of Mounier's essays of which the most important for my purposes was first published in 1948 under the appropriate title, "La Petite Peur Du XXe Siecle."

[9] *Ibid.*, p. 30.

[10] *Ibid.*, pp. 33, 49.

[11] *Ibid.*, p. 44.

[12] *Ibid.*, p. 44. Cox recognizes this myth as an ancient one and rightly argues that Judaeo-Christian notions of Creation contain a disenchantment with nature romanticism. However, Cox's acceptance of a neo-positivism in van Peursen's redoing of Compte's three ages (p. 64, *The Secular City*) perhaps prevents him from seeing that myths tend to outlive any single form in their history. Here Ricoeur's profound study, *La Symbolique du mal* (Paris: Aubier, 1960) is instructive.

[13] *Ibid.*, p. 56.

[14] *Ibid.*, p. 56.

[15] *Ibid.*, p. 22.

[16] *Ibid.*, p. 21.

[17] *Ibid.*, p. 17.

[18] *Ibid.*, p. 47.

[19] *Ibid.*, p. 64.

[20] *Ibid.*, p. 50.

[21] *Ibid.*, p. 21.

[22] If one reads both Mounier and Berdyaev he should see that an extravagant individualism is lacking there too. Berdyaev, after all, elaborated upon the idea of *sobornost*, and Mounier's personalism was always also a communalism.

[23] Ricoeur's recently translated essays contained in *History and Truth* (Evanston: Northwestern University Press, 1965) were originally published in the fifties.

[24] Paul Ricoeur, *History and Truth*, trans. by Charles Kelbley (Evanston: Northwestern University Press, 1965), p. 82.

[25] *Ibid.*, p. 83.

[26] *Ibid.*, p. 85.

[27] *Ibid.*, p. 190.

[28] *Ibid.*, p. 107.

[29] *Ibid.*, p. 103.

[30] *Ibid.*, p. 106.

[31] *Ibid.*, p. 105.

[32] *Ibid.*, p. 172.

INDEX OF NAMES

Abbot, Edwin 81n.
Aristotle xxi, 82

Bach, J. S. xxvii, 94, 96, 99
Bacon, Francis 134
Berdaev, Nicolas xxv, 142

Camus, Albert 141
Cohen, Robert x
Cowan, Ruth ix
Cox, Harvey 141–143, 154–146, 148–159

de Beauvoir, Simone 141
Democritus xx, 38
Dostoyevski, Fydor xxv
Dreyfus, Hubert 58

Faustus, Dr. xix
Frankenstein xix, 40

Goedeke, Walter R. 82

Heaton, J. M. 86
Heelan, Patrick ix
Heidegger, Martin xxi, xv–xxvi, 4, 7, 19, 23, 28, 39, 48, 65–66, 80n., 103, 132, 142
Heraclitus 82
Husserl, Edmund xvii–xviii, 3, 5, 15, 66–67, 117–118

Jaspers, Karl xxv, 142
Jonas, Hans xxvii, 130, 133–139

Koffka, Kurt 87
Kuhn, Thomas xxv

Luther, Martin 97

Marcel, Gabriel xxv
Marcuse, Herbert 131
Merleau-Ponty, Maurice 7, 15, 19, 85, 87, 88, 90–91, 99, 141, 142
Miller, Lee ix
Messing, Robert 100n.
Mournier, Emmanuel xxvii, 58, 130, 132, 133–139, 142–146

Ong, Walter 90

Piaget, Jean 87
Plato xix–xx
Pourrat, Henri 144

Ricoeur, Paul xxvii, 142–143, 146–148

Sartre, Jean Paul xvii, 141–142
Sellars, Wilfred 23
Spector, Marshall ix
St. Exupery 144

Truxall, John 58

Vivaldi, Antonio 94, 99

Wartofsky, Marx x
White, Lynn xix
Wittgenstein, Ludwig xviii